Technical Career Survival Handbook

Technical Career Survival Handbook

100 Things You Need to Know

Peter Y. Burke P.E.

AMSTERDAM • BOSTON • HEIDELBERG • LONDON
NEW YORK • OXFORD • PARIS • SAN DIEGO
SAN FRANCISCO • SINGAPORE • SYDNEY • TOKYO
Academic Press is an imprint of Elsevier

Academic Press is an imprint of Elsevier
125 London Wall, London EC2Y 5AS, United Kingdom
525 B Street, Suite 1800, San Diego, CA 92101-4495, United States
50 Hampshire Street, 5th Floor, Cambridge, MA 02139, United States
The Boulevard, Langford Lane, Kidlington, Oxford OX5 1GB, United Kingdom

Notices
Knowledge and best practice in this field are constantly changing. As new research and experience
broaden our understanding, changes in research methods, professional practices, or medical treatment
may become necessary.

Practitioners and researchers must always rely on their own experience and knowledge in evaluating and
using any information, methods, compounds, or experiments described herein. In using such information
or methods they should be mindful of their own safety and the safety of others, including parties for
whom they have a professional responsibility.

To the fullest extent of the law, neither the Publisher nor the authors, contributors, or editors, assume any
liability for any injury and/or damage to persons or property as a matter of products liability, negligence
or otherwise, or from any use or operation of any methods, products, instructions, or ideas contained in
the material herein.

Library of Congress Cataloging-in-Publication Data
A catalog record for this book is available from the Library of Congress

British Library Cataloguing-in-Publication Data
A catalogue record for this book is available from the British Library

ISBN: 978-0-12-809372-6

For information on all Academic Press publications
visit our website at https://www.elsevier.com/

Working together
to grow libraries in
developing countries

www.elsevier.com • www.bookaid.org

Publisher: Sara Tenney
Acquisition Editor: Mary Preap
Editorial Project Manager: Mary Preap
Production Project Manager: Chris Wortley
Designer: Matthew Limbert

Typeset by TNQ Books and Journals

Transferred to Digital Printing in 2017

Contents

Acknowledgments

I give thanks to Jesus Christ, my Lord and savior for his steadfast love and gift of perseverance. This has allowed me to pursue and survive a challenging and rewarding technical career.

I must acknowledge my lovely wife Ellen who was upset about my insistence on monopolizing our PC rather than my laptop to prepare this book. I appreciate her patience and understanding throughout the process.

There are several friends whom I would like to convey my appreciation for their reviews, comments, and suggestions for this book. Thank you Frank Brooks, Don Kinniburgh, Bill Rhoney, Steve Rodimer, Jack Smith, and Carl Wise for their contributions.

Finally, to my Elsevier publicist Mary Preap and her support staff, thank you for selecting my manuscript and supporting me throughout the publication process.

Introduction

Congratulations! You have taken a big step toward gaining insight on your chosen career. Whether you are in the planning stage or active in the workforce, there is always a need to gain information about what lies ahead. That is exactly what this book is about. Some would question why it is about things you need to know to survive and not about "the key to success." To be successful, you must first survive all the situations you will be faced with, many of which may not be attributed to your knowledge, ability, or drive. Does not everyone want to survive? You will need to make good decisions based on a collection of the facts. I believe this book will help you in that regard. Whereas to succeed, you will need to continuously increase your knowledge of your chosen field and of course, work hard and smart. You are on your own there.

A technical career can be a very rewarding vocation and provide great satisfaction whether you are part of a large team or an individual consultant to industry. However, it will not make you the chief conversationalist at the neighborhood cocktail parties. Reverend Brian Fletcher, a friend of mine claims he can terminate a conversation quicker than me by mentioning his profession. It is quite difficult for nontechnical individuals to understand and appreciate your challenges unlike those of a teacher, nurse, or a rock star. This is probably because the term "engineer" is so often misused in our society. Two classic examples: "domestic engineer" is often used to describe a person who manages a household or the "engineer" is used to refer to a person who operates a railroad engine.

By reading this book, you will gain insight on the business world based on my experiences. It became apparent to me that my career was much more diverse than many others in my field. Working in locations in the North, South, East, and West for small, medium, and large corporations, self-employed, private and public companies, service, utilities, and manufacturers have all provided me with a wealth of knowledge regarding technical career situations. Now, please realize that I did not set out to average 5 years per employer. Neither did my three engineer friends at Northrop Grumman, General Electric, and Dominion Virginia Power Nuclear, who set out to work for one company for their entire careers. Rather, we made decisions over the years that kept us on our respective career paths. You too will encounter situations requiring prudent decisions along *your* path as well. But I am quite certain, the information in this book will better equip you to deal with those tough decisions. I suggest you read the entire book, then later, as the occasions arise, refer to specific subjects herein as a how to guide.

This book *is not* an autobiography and I did not set out to dwell on all the details and ups and downs of my career. However, I did refer to situations that I faced to equip you for the challenges *you* will face. Also bear in mind, my perspective is strictly from a mechanical engineering standpoint. However, most of the subject matter in this book is relevant to many of the disciplines in engineering, science, and technology.

To provide continuity, the subjects in this book are arranged roughly in the chronological order that you can expect to be confronted with them but not necessarily. Surprisingly, "the 100 things" came to me rather quickly. You will see that some subjects are far more complex than others but not necessarily more or less important. Additionally, with an ever-changing economic and political environment, you may encounter circumstances drastically different than mine. Perhaps those circumstances will even be sufficient for a book of your own.

Oh yes, regarding the song title references, I pulled some of them from the many songs in my classic rock and roll collection. Playing guitar and singing, sometimes for charity events, has always been one of my favorite diversions from the rigors of a technical career. Thankfully, I did not have to depend upon my passion for music for a living. Believe me, you will need diversions too! But be careful not to have too many.

Song: "I Can See Clearly Now"
As popularized by: Johnny Nash
1972

Part I

Preparing for the Work Force

Technical Careers

"Technical" tends to be a broad term that is generally related to principles of some science. So let us clarify the term "technical careers." Does it specifically mean engineers, scientists, or teachers? For the purpose of this book, I am referring mainly to **engineers** and graduates with technical degrees that typically work in an organizational team toward a common objective. I do not include teachers or professors in this definition even though some may be considered a part of the scientific community. Their primary focus is obviously teaching and research within the field of education.

I do include certain **scientists** in the scope of this book, particularly those that are entrepreneurial and make innovation happen. They blend their scientific knowledge and credibility with people skills, entrepreneurial ship, and business knowledge. They most always have a PhD and spend more time for analyzing things in broad strokes than engineers who spend time in building things. I have experience working with them as part of an engineering team albeit limited. Computer scientists often work in an innovative organizations, many high tech industries. They are like many other disciplines that the Bureau of Labor Statistics (BLS) categorizes such as biochemists, biophysicists, chemists, material scientists, environment scientists, and physicists.

Regarding computer *support* personnel, they are often positioned as staff personnel and cover the needs of *several* "teams," for that reason, they may not be within the scope of this book. They typically do not participate in a product or process development team but provide IT help when called upon.

I also refer to **technicians** who may have some academic background, possible job training and profess a knowledge of certain machinery, electronics, structures, materials, test equipment, processes, and instrumentation. Many work alongside engineers and assist with prototype construction, performance testing, and design modifications. Often they obtain advanced degrees to become engineers. Many gained experience while serving in the military with equipment and systems (Table 1.1).

While I was director of engineering at Sundstrand Fluid Handling (SFH), Howard Ammons was the lead technician who reported to me and supervised our test lab. He could make a rough sketch of a precision part on the back of an envelope, machine it and test it in a pump in practically the same amount of time it would take a designer to produce a computer-aided design production drawing.

Technical personnel discussed in this book, apply the principles of science and mathematics to develop practical solutions to technical problems. Often their work involves translating scientific discoveries into commercial applications thereby meeting societal and consumer needs also known as "the market."

Technical Career Survival Handbook. http://dx.doi.org/10.1016/B978-0-12-809372-6.00001-3

Table 1.1 **Major Technician Occupations**

Description	2012 US Employment Actual	2022 US Employment Projected
Aerospace technicians	9900	9900
Civil technicians	73,100	73,600
Electrical/electronic technicians	146,500	146,500
Electromechanical technicians	17,300	18,000
Environmental technicians	19,000	22,500
Industrial technicians	68,000	65,800
Mechanical technicians	47,500	49,700

BLS 2014 data.

Their work may involve developing totally new products, product derivatives, correcting product deficiencies, or installing equipment. They often evaluate performance, cost, size, efficiency, safety, reliability, and manufacturability. Generally, technical personnel are dedicated to a specific product or service so that their knowledge is focused and specialized. Areas of specialization may include chemical production, computers, power plants, aircraft, pumps, compressors, solar panels, and even toys.

In years past, engineers, assisted by designer/drafters, typically laid out designs on paper before prototypes were built, but with the advancement of central processing unit chips, including Intel's latest I7 quad core processor, today's computers are used to analyze and produce designs; to simulate and test how a machine, structure, or process operates; to generate specifications for parts; to monitor the quality of products; and to control the efficiency of processes.

Song: "Analog Man"
As popularized by: Joe Walsh
2012

The Big Four Engineering Degrees

2

Chapter Outline

Let us look at some of the various disciplines of engineering study and briefly explain their functions starting with the big four. These curriculums the largest segment of the engineering world and represent 60% of all engineering disciplines. There are more positions occupied by graduates with these degrees than any others. Logically, there are more graduates obtaining degrees in these curriculum than the others. Also, some might claim that these four degrees are the most versatile and therefore one would be more likely to survive in the job market. These might be the factors that influence the decision as to which engineering degree to obtain but it is also important that you pursue a degree in the field you are mainly interested. Also bear in mind that other curriculums such as biomedical and petrochemical engineering may offer higher starting salaries. But ironically at this moment, oil companies are laying off thousands of petroleum engineers due to the plentiful supply of oil.

Civil Engineering

Civil engineers design and supervise projects like the construction of roads, buildings, airports, tunnels, dams, bridges, and water supply and sewage systems. They consider many factors in the design process such as construction costs, expected lifetime of a project to government regulations, and potential environmental hazards including earthquakes and hurricanes. Civil engineering, considered one of the oldest engineering disciplines, encompasses many specialties. The major ones are structural, water resources, building construction, transportation systems, and geotechnical engineering. Civil engineers who specialize in structural analysis are often referred to as civil/structural engineers.

Electrical Engineering

Electrical engineers design, develop, test, and supervise the manufacture of different electrical equipment. This equipment includes electric motors; machinery controls, lighting, and wiring in buildings; radar and navigation systems; communications

Technical Career Survival Handbook. http://dx.doi.org/10.1016/B978-0-12-809372-6.00002-5

systems; and power generation, control, and transmission devices used by electric utilities. Electrical engineers also design the electrical systems contained in automobiles and aircraft. Although the terms *electrical* and *electronics engineering* often are used interchangeably, electrical engineers traditionally have focused on the generation and supply of power, whereas electronics engineers work on applications of electricity to control systems or signal processing. Some electrical engineers specialize in areas such as power systems engineering or electrical equipment manufacturing.

Mechanical Engineering

Mechanical engineers conduct research, design, develop, produce, and test tools, engines, machines, and other mechanical devices. Mechanical engineering is one of the broadest and largest engineering disciplines. Engineers in this discipline may choose to work on power-producing machines such as electric generators, internal combustion engines, wind power, solar power, fuel cells, and steam and gas turbines. They may also work on power-using machines such as refrigeration and air-conditioning equipment, machine tools, material-handling systems, elevators and escalators, industrial production equipment, and machinery used in manufacturing. Some mechanical engineers design tools that other engineers use. In addition, mechanical engineers work in manufacturing or agriculture production, maintenance, or technical sales; many become supervisors or managers.

Industrial Engineering

Industrial engineers determine the most effective ways to use the basic production principals, people, machines, materials, information, and energy to make a product or improve the performance of a service. They are concerned principally with increasing productivity through personnel management, methods of business organization, and technology. To maximize efficiency, industrial engineers study product requirements carefully and then design manufacturing and information systems to meet those requirements using mathematical methods and models. They develop management control systems to aid in financial planning and cost analysis, and design production planning and control systems ensure product quality. They also design or improve systems for the physical distribution of goods and services and optimize plant locations.

Song: "Cat's in the Cradle"
As popularized by: Harry Chapin
1974

Other Engineering Degrees

Alternatives to the big four described previously are included in this chapter. They are by no means less challenging or more readily obtainable. They are simply less common and may not be offered at all technical colleges and universities.

Aerospace engineers design, test, and supervise the production of aircraft, spacecraft, and missiles. *Aeronautical engineers* work with aircraft and *astronautical engineers* work on spacecraft. Aerospace engineers develop new technologies for use in aviation, defense systems, and space exploration, and specialize in areas such as structural design, guidance, navigation, and controls. They also may specialize in aerospace products, such as commercial aircraft, military aircraft, helicopters, spacecraft, or missiles and rockets. They may exhibit expertise in aerodynamics, thermodynamics, propulsion, acoustics, or guidance and instrumentation systems.

Agricultural engineers apply their knowledge of engineering technology and science to agriculture and the efficient use of biological resources. Sometimes they are referred to as *biological engineers*. Some design agricultural machinery, equipment, sensors, processes, and structures used for crop storage. Some specialize in areas such as power systems and machinery design, structural and environmental engineering, and bioprocess engineering. They may develop ways to conserve soil and water and to improve the processing of agricultural products. Agricultural engineers often work in fields such as research and development, production, sales, or management.

Biomedical engineers develop procedures and devices that solve medical and health-related problems combining biology and medicine with engineering principles. Many specialize in research, along with medical scientists and develop and evaluate systems and products such as artificial organs, prostheses, instrumentation, medical information systems, and health management and care delivery systems. Biomedical engineers may also design devices used in various medical procedures, imaging systems such as magnetic resonance imaging, and devices for controlling body functions.

Chemical engineers apply the principles of chemistry to solve problems involving the production or use of chemicals and other products. They design processes and equipment for large-scale chemical manufacturing, plan and test methods of manufacturing products and treating byproducts, and supervise chemical production. Chemical engineers also work in different manufacturing industries other than chemical manufacturing, such as those producing energy, electronics, paper, food, and clothing. Some work in health care, biotechnology, pharmacology, and various business services. Chemical engineers also apply principles of physics, mathematics, and mechanical and electrical engineering, as well as chemistry.

Computer engineers research, design, develop, test, and oversee the manufacture and installation of computer hardware such as computer chips, circuit boards,

Technical Career Survival Handbook. http://dx.doi.org/10.1016/B978-0-12-809372-6.00003-7

computer systems, and related equipment/components. Computer software engineers often called computer engineers design and develop the software systems that control computers. The work of computer hardware engineers is similar to that of electronics engineers in that they may design and test circuits and other electronic components only as they relate to computers and ancillary equipment.

Electronics engineers are responsible for technologies such as portable music players and global positioning systems, which can provide the location of, for example, a vehicle. Electronics engineers design, develop, test, and supervise the production of electronic equipment such as broadcast and communications systems. Many electronics engineers work in areas closely related to computers. However, their work is related exclusively to computer hardware and they are considered computer hardware engineers. Electronics engineers specialize in areas such as communications, signal processing, and control systems.

Engineering science and mechanics engineers use fundamental principles to develop engineering solutions to contemporary problems in the physical and life sciences. Their work also involves fluid mechanics, dynamics and vibration, biomechanics, and computational methods. Many of the areas of study under this field are also included in the curriculum of other fields mainly civil and mechanical engineering.

Environmental engineers use principles of biology and chemistry to develop solutions to environmental problems. Their work involves water and air pollution control, recycling, waste disposal, and various public health issues. Environmental engineers conduct hazardous waste management studies and evaluate the significance of the hazard, provide recommendations about treatment and containment, and develop regulations. They participate in the design of municipal water supply and industrial wastewater treatment systems, conduct research on the environmental impact of proposed construction projects, analyze scientific data, and perform quality control procedures. Environmental engineers may be concerned with local and worldwide environmental issues. Some may attempt to minimize the effects of acid rain, climate change, automobile emissions, ozone depletion, and the protection of wildlife.

Fire protection engineering is offered at a few colleges with emphasis on protecting people, property, and environments from the harmful effect of fire and smoke. The focus is on detection, suppression and mitigation, and personnel safety. Fire protection engineers identify risks and design safeguards that aid in preventing, controlling, and mitigating the effects of fires. They may assist design engineers, architects, and building owners and developers in evaluating a buildings life safety and property protection goals.

Green engineering according to Environmental Protection Agency (EPA) is the design, commercialization, and use of processes and products, which are feasible and economical while first minimizing the generation of pollutants at the sources and second the risks to human health and to the environment. This study provides tools for technologists to assess the impact of their processes and systems that affects air, water, and soil. Because this is a relatively new field of study, only a limited number of colleges, universities, and institutes currently offer it as either a minor or major. The University of Central Florida's Solar Energy Center offers a mechanical engineering degree with a "specialization" in solar.

Health and safety engineers, except mining safety engineers and inspectors, prevent personnel injury and property damage by applying their knowledge of systems engineering and mechanical, chemical, and human performance principles. Using this specialized knowledge, they identify and measure potential hazards, such as the risk of fires or handling toxic chemicals. They recommend appropriate loss prevention measures based on their probability of harm and potential damage. They develop procedures and designs to minimize the risk of illness, injury, or damage. Some work in manufacturing industries to ensure that the designs of new products do not create unnecessary hazards and evaluate hazardous conditions, as well as develop hazard control methods.

Manufacturing engineers are involved in all aspects of manufacturing products. They design everything from the mechanical and electrical components of the products, to automated assembly processes, to the supply chain that gets materials to the factory. Companies must produce their products at a competitive cost to be profitable, so the manufacturing engineer must have an understanding of business and economics in addition to hands-on technical skills. Manufacturing engineering is becoming increasingly sophisticated: researchers in the field work to develop such equipment as fuel cells, robotic systems, green manufacturing processes, and microelectromechanical devices and systems.

Marine engineers and naval architects are involved in the design, construction, and maintenance of ships, boats, and associated equipment. They design and manage the construction of everything from aircraft carriers to submarines and from sailboats to tankers. Naval architects work on the basic design of ships, including hull stability. Marine engineers work on the propulsion, steering, and other ship systems. Marine engineers and naval architects apply knowledge from a range of fields to the entire process by which water vehicles are designed and manufactured. Workers who operate or supervise the operation of marine machinery on ships and other vessels are often referred to as marine engineers or ship engineers.

Materials (metallurgical) engineers are involved in the development, processing, and testing of the materials used to manufacture products from computer chips and aircraft wings to golf clubs and snow skis. They work with metals, ceramics, plastics, semiconductors, and composites to create new materials that meet specific mechanical, electrical, and chemical tolerances and specifications. They may be involved in selecting materials for new applications and develop the ability to create and then study materials at an atomic level, using advanced processes. Most materials engineers specialize in a particular type of material. For example, metals such as steel, ceramic materials, and the processes for making them into useful products.

Mining and geological engineers find, extract, and prepare coal, metals, and minerals for use by utilities and manufacturing industries. They design open pit and underground mines, supervise the construction of mines and tunnels in underground operations, and devise methods for transporting minerals. They are responsible for the safe, economical, and environmentally sound operation of mines. Mining engineers often work with geologists and metallurgical engineers to locate and appraise new ore deposits. Some develop new mining equipment or direct mineral-processing operations that separate minerals from the dirt, rock, and other materials. Mining engineers frequently specialize in the mining of a particular mineral or metal, such as coal or gold. Many are working to solve problems related to land reclamation, water and air pollution, and protection of the environment.

Nuclear engineers research and develop the processes, instruments, and systems used to derive benefits from nuclear energy and radiation. They design, develop, monitor, and operate nuclear plants to generate power. They may work on the nuclear fuel cycle—the production, handling, and use of nuclear fuel and the safe disposal of waste produced by the generation of nuclear energy or on the development of fusion energy. Some specialize in the development of nuclear power sources for naval vessels or spacecraft or nuclear weapons while others find industrial and medical applications for radioactive materials, i.e., in equipment used to diagnose and treat medical conditions.

Petroleum engineers design procedures for extracting oil and gas from deposits found beneath the earth or ocean. Petroleum engineers work with geologists and other specialists to understand the formation and properties of the rock containing the reservoir and determine the drilling methods and to monitor drilling and production operations. They design equipment and processes to achieve the maximum recovery of oil and gas. Because only a small proportion of oil and gas in a reservoir flows out under natural forces, petroleum engineers use various enhanced recovery methods, water injection, chemicals, gases, or steam into an oil reservoir to force more oil from the earth.

Engineering degrees are typically earned in about 4–5 years. There is a significant overlap of courses from one engineering curricula to another. For example, during the first or second year, courses such as math, English, chemistry, and physics are required for most engineering programs. Similarly, in the second and third years, electrical engineering (EE) courses may be required for mechanical engineering (ME) degrees and conversely some ME courses (typically thermodynamics) will be required for an EE degree. Certain EE and ME courses are common among most of the engineering programs listed earlier.

I should point out here that it is not unusual for an engineering major to switch fields of study after being exposed to courses within a specific engineering discipline. This may occur as the student becomes more familiar with specific course requirements. For example, a chemical engineering major may decide that mechanical or civil engineering would be more favorable based on difficulties grasping the fundamentals of chemistry. Then there is a college friend, Bruce Davey who foresaw that his major, chemical engineering was too challenging. So he switched his major to biology and later graduated from medical school and became a successful orthopedic doctor. Realize of course that the switch may be at the expense of time, tuition money, and catching up with those that remained on a constant path. But it is better for survival to make a career path correction early rather than late in one's academic studies.

In 2013, engineers held about 1.5 million jobs with US-based employers. These data would indicate that fields such as civil and mechanical have more job opportunities than, say, biomedical or nuclear engineering if you are concerned about employment opportunities after receiving your degree. This is only one factor to consider. More factors will be discussed in Part II of this book.

Conclusions from the 2008 BLS data are as follows (Table 3.1):

Figure 3.1 Construction of a typical power plant requires multiple disciplines.

Table 3.1 **Distribution of Employment by Engineering Discipline**

Engineering Discipline	Number Employed
Civil engineers	270,000
Mechanical engineers	260,000
Electrical engineers	235,000
Industrial engineers	225,000
Electronics engineers	135,000
Computer hardware engineers	85,000
Aerospace engineers	75,000
Environment engineers	54,300
Petroleum engineers	35,000
Chemical engineer	33,000
Health and safety engineers	24,000
Materials engineers	24,000
Biomedical engineers	24,000
Marine engineers	20,000
Nuclear engineers	16,900
Mining and geological engineers	8000
Agricultural engineers	3000

BLS Data (2013).

- Approximately 36% of the jobs were found in manufacturing industries.
- Approximately 30% were found in professional, scientific, and technical services (consulting) industries.
- Approximately 6% of engineers are employed in state and local government.
- Approximately 6% of engineers are employed in federal government.
- Only about 3% of engineers are self-employed as consultants.

Song: "Land of Confusion"
As popularized by: Genesis
1986

Technology Degrees

4

In addition to technical degrees (Bachelor of Science and so on), which are generally 4-year programs, graduates with technology degrees also pursue technical careers. These degrees can typically be obtained in 2 years, often at community colleges or online programs. I include computer-aided design (CAD) personnel who are often referred to as "designers." Many have a few years of engineering study, associate degrees or focus on specific courses in software usage such as CAD and modeling. Consequently, they often work side-by-side with engineers although many engineers are also trained and utilize CAD as well (Fig. 4.1).

Today there are many personnel who have military experience or a few years of on-the-job training that are in the work force engaged in technical careers. However, they will likely experience difficulty in climbing the corporate ladder today compared with past years due to competition with a well-educated, younger work force.

Several typical technology programs include the following:

Information technology—focuses on project management, systems development, networking, programming language, and Internet concepts and design.

Information technology networking—focuses on networking aspects of IT, including local area networking and wide area networking.

Civil engineering technology—land surveying, technical writing, mathematics, and construction and design.

Mechanical engineering technology—is the application of physical principles and current technological developments to the creation of useful machinery and operation design. Through the application of computer-aided manufacturing, models may be used directly by software to create "instructions" for the manufacture of objects represented by the models through computer numerically controlled machining or other automated processes.

Electrical engineering technology—focus on design, drafting, and technical skills they need to assist engineers and implement engineering ideas. Classes cover such topics as electrical circuitry, systems analysis and testing, and instrument preparation.

Manufacturing engineering technology—focus on automated manufacturing and materials handling using computers to design and manufacture products or in process and quality control. Students learn to take individual islands of manufacturing automation and link them together to achieve the ultimate efficiency of an organization's manufacturing resources.

If your interest lies in pursuing a technology degree, you may want to contact the college you are considering to obtain data pertaining to the number of workers in the US work force possessing the degree you are considering. You should also inquire as to the job demand for graduates with this degree.

Technical Career Survival Handbook. http://dx.doi.org/10.1016/B978-0-12-809372-6.00004-9

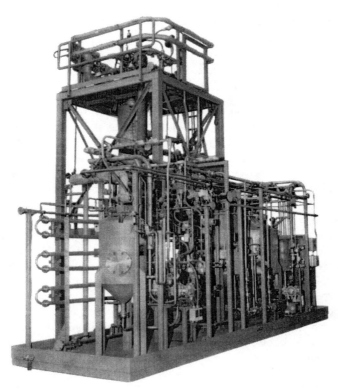

Figure 4.1 CAD drawing of a chemical process skid.

On a note of clarification, the term "technology" has been somewhat hijacked in recent years and is often referred to in conjunction with Internet technology (IT), computers and related product manufacturing. Hereafter, I will be referring to technology in a more general sense relating to products and services that are highly engineered regardless of the industry.

Song: "Wonderful World"
As popularized by: Sam Cooke
1957

Career Choice

Chapter Outline

The cost of obtaining a degree has skyrocketed and is perhaps not a practical investment for all college eligible students. Colleges have increased tuition dramatically because, well, because they can with little or no consequences. Meanwhile, demand exceeds supply and college enrollments just continue to increase. Also, readily available government-backed student loans have eased the pressure on colleges to cut costs. So, it is imperative that the decision to pursue a degree be well-thought-out. If you are contemplating a technical career, here are four questions to consider, which may help your decision-making process.

What Are My Interests?

My brother Jim and I had distinctly different interests, which provided an obvious clue as to what fields of study we were to pursue. Jim enjoyed stringing wire to a neighbor's barn setting up telegraph and phone communications systems, and operating ham radio equipment. I, on the other hand, built a go-cart from a lawn mower handle and engine and tore apart old Ford cars and modified the engines, bodies, and transmissions. It became pretty obvious that Jim was to become an electrical engineer and I was to become a mechanical engineer. At least that is the way we saw it.

Unfortunately for some, it may not be as obvious. You may not have any interest in getting your hands on a 5.0 L Ford V-8 engine and tearing apart the four-barrel carburetor. That does not mean you should not consider studying mechanical engineering and that you cannot be a competent engineer. One of my friends and previous coworker, Bill Rhoney, had a love for both math and piano while in middle and high school. However, he realized that teaching music within a high school setting was not his top choice of vocation and came with limited earnings potential. He recognized that math would integrate well within an engineering program and went on to earn his masters degree in mechanical engineering and had a successful career with a major chemical corporation.

Technical Career Survival Handbook. http://dx.doi.org/10.1016/B978-0-12-809372-6.00005-0

So, a curiosity of how and why things work the way they do is typically an attribute found in a lot of budding engineers. Perhaps structures such as bridges, stadiums, or high rise buildings fascinate you. This might be a sign that you are interested in studying civil engineering. Interested in marine life or the environment? Find out what engineering opportunities exist in those areas.

What Subjects Do I Like?

Naturally, math is an important subject at the high school level for those contemplating engineering studies. I would also recommend evaluating what aspects of physics and chemistry you find appealing, particularly the former. High school physics deals with many of the aspects of engineering, i.e., chemical, electrical, mechanical, aeronautical, nuclear, and more. Chemistry and mathematics are key subjects for chemical engineers. So take the opportunity to judge your interest from a subject perspective.

What Subjects Are My Strengths?

Often your likes are your strengths. But do not necessarily take that for granted. The technical subjects you excel in are a good indicator that you have the ability to grasp technical concepts be they a mathematical, physical, or chemical nature. Whereas strength in artful subjects, literature, or history would suggest a different career path.

Who Do I Know in the Field?

Knowing someone working in the field that you are considering is a great way to become familiar with what the work entails. Do your parents work in the field, cousins, neighbors, aunts, uncles, or siblings that work in the field? If so, spend time with them, ask questions, and/or possibly spend some time with them on their job site or in their office or shop.

I was fortunate to spend two of my college summers working as a gopher/clerk in the office of Devenco, Inc., a consulting engineering company located in New York City, although I probably spent the bulk of my earnings commuting from my home in New Jersey. My work was very basic and consisted of chasing down parts, obtaining technical publications (precomputer era) and building a technical library to support the office staff, many of whom were working on classified government projects.

During my summers there, I gained a good feel for the world of engineering and how engineers spent their days. One of the R&D engineers nicknamed "Tuna" worked on a device we played with in the lab during lunch hour 1 day, but I did not appreciate it until several years later. During the notorious Watergate scandal, President Richard Nixon used an electronic, voice-actuated, 7 s delayed tape recorder that could be used to record secret telephone conversations. When that story later made the news, I realized that Tuna's project helped make a piece of history. Today, electronic voice actuators practically turn on instantaneously.

Beware, however, that none of the above questions alone can ensure survival in your chosen technical field. There are many books available that delve deeply into career choices. Spend time in researching the fields that appeal to you. Particularly whether the end product of your work appeals to you. For example, you might realize that building jet aircraft engines is a way more appealing to you than designing and building automation equipment to make plastic film or glass bottles on a production line. You may also want to consider aptitude tests to narrow your choice of fields. It is unfortunate that little time is spent in understanding what a particular career encompasses until later when we are fully committed.

My brother Jim made a major career change several years after graduating with an electrical engineering degree, upon discovering he was better suited for accounting. Aptitude tests pointed toward strength in ability of working with numbers versus engineering principals. Unfortunately, this was discovered after considerable time and money had been invested in an engineering career path much to the dismay of our parents. Lesson learned, make a serious effort to determine what career path is most beneficial for you.

Song: "Maggie May"
As popularized by: Rod Stewart
1971

College Selection

This is the subject of many dinner conversations between parents and their children starting during their junior year of high school. It is an experience, which parents learn how much influence they have had while raising their children and how realistic their children's perception of the real world is. In my case, my brother graduated from Virginia Tech during my junior year in high school. So, of course, I figured if my brother could do it, then how hard could it be? Very!

I will not attempt to cover many of the topics that are readily available in books, magazine articles, and reports in the public domain. I will simply relate my experience in the selection process. Of course, I weighed the main factors of cost, location, college ranking in the field of study, and college entrance requirements as well as any youth can be expected. But it was not until I was working in my first job as a recent hire that I experienced the college "reputation" factor head-on.

Typically, college juniors or seniors sign up for on-campus interviews with recruiters from human resource departments of various companies seeking to find new employees. Assuming the on-campus interview is satisfactory, the expectant graduates are later invited to travel to the employer's location for a facility tour and more detailed interviews.

Hamilton Standard Division (HSD) of United Technologies (UTC) frequently relied on the newbies to host and usher the perspective engineering recruits around the plant and offices, line up their interviews, show them the production facilities, and take them to lunch. As a host, I encountered some reluctance on the part of department managers to interview certain recruits based on the college where they would soon matriculate. This emphasized to me the importance of people place on the reputation of colleges and universities.

So what does this imply from a practical standpoint? The 2012 U.S. News and World report ranking for 198 engineering universities evaluates factors such as student faculty ratios, research expenditures, peer assessment, recruiter assessment, mean Graduate Record Exams (GRE) scores and several other parameters. (Refer to the complete report for details.) There are 42 states represented in the top 100 engineering universities.

Interestingly the states with the most engineering universities are:

California with 11
Texas with 7
New York with 7
Massachusetts with 6
Pennsylvania with 5

So we can conclude there is likely a highly rated university in your state albeit not necessarily in your hometown. Hence you may be able to avoid the additional expense

Technical Career Survival Handbook. http://dx.doi.org/10.1016/B978-0-12-809372-6.00006-2

associated with out-of-state tuition. You may also want to check the college acceptance rate as a percentage of applicants that are actually admitted. According to Money Magazine, August 2015, Texas A&M University and Virginia Tech were ranked at 69% and 70% acceptance ratings. This is surprisingly high when compared to many colleges with single digit ratings.

So the bottom line is this, target a top college and your likelihood of landing a satisfying job upon graduation will be increased. From a practical standpoint, it may be necessary to attend a lower ranked college or community college for your first couple years and then transfer into a higher ranked school to complete your studies. Often transferring to a college is less competitive academically than entering as a freshman particularly if your high school standing is not exceptional. It will be well worth the effort in the end. Certainly, cost and location are major considerations but, in any case, aim high.

Interestingly, Virginia has 23 community colleges according to the State Council of Higher Education for Virginia (SCHEV). They recently concluded that an in-state student can save $15,000 or 30% tuition and fees for a BS degree by attending a community college for the first 2 years. Oh yes, those "fees" are becoming a major portion of the yearly college expense. In some instances, as much as 28% of the total annual expense.

Song: "Kodachrome"
As popularized by: Paul Simon
1973

Engineering Technology

Although closely related, engineering technology is often confused with engineering. The former focus is on application and implementation requiring courses of study such as algebra, trigonometry, applied calculus, and courses of a more practical nature rather than theoretical. Whereas the latter involves theory and conceptual design with courses of study including multiple semesters of calculus, engineering mechanics, fluid dynamics, etc.

Students who graduate from an accredited 4-year engineering program are called "engineers" and awarded a Bachelor of Science degree. Some proceed to further their education and pursue advanced degrees such as a Master of Science or Doctorate in engineering. Those students who obtain a 4-year technology degree should be referred to as "technologists" although in many companies they are inadvertently called "engineers" and conform to the engineer job descriptions. Many enter the workforce in sectors such as manufacturing, construction, product design, computer science, sales, or applications/marketing positions.

Two-year engineering technology programs generally result in an Associate of Science (A.S.) degree and graduates are referred to as "technicians." Four year programs may require up to 130 credit hours much like engineering programs. These students receive a Bachelor of Science in Technology. They may enter similar fields as those mentioned for "technologists" or proceed with obtaining a 4-year engineering degree. One institution offers a masters in Mechanical Technology based on completion of 30 credit hours plus project work.

However, these students must be very cautious and understand the requirements of the 4-year institution they are considering as a transfer of course credits is not automatic by any means.

Some degrees are awarded to students that have undertaken a degree program, which is additionally supplemented by either occupational placements (e.g., supervised practice or internships), practice-based classroom courses, or associate degrees. Due to these requirements, the degree may take at least 4 years.

This is a subject I learned about through first-hand on-the-job (OTJ) experience. Like many situations you face on the job, they are not the subject of a course you studied in college, i.e., hiring. While I was managing a development engineering department at Sundstrand Fluid Handling we were expanding our engineering ability to bring new products to the market. I had budget approval to add an entry level engineer to work alongside a senior engineer to evaluate adapting an existing centrifugal pump line to power recovery turbine service. Think of a pump used in reverse. A pressurized fluid enters the turbine and causes it to rotate an impeller and drive a device such as a generator on the other end of the shaft assembly.

Technical Career Survival Handbook. http://dx.doi.org/10.1016/B978-0-12-809372-6.00007-4

We interviewed several candidates for the position and finally settled on a recent graduate with a "technology degree." Initially, he seemed to fit into the group well and appeared to be conscientious regarding his work output. But slowly it became apparent that he did not have an understanding of the cause and effect and was struggling with the fundamental principles of the equipment design. In development work, it is important to explain results of tests even when they do not produce expected results. Then you make adjustments/changes to the design and repeat. It takes a certain curiosity, which he did not possess. I took a closer look at his college curriculum and realized that most of his courses were a practical treatment of the subject as opposed to theoretical in nature. We finally had to let him go.

However, in other less theoretical job situations, the technology degree may have been adequate for other positions we had available. For example, some institutions offer career opportunities that may be available for individuals with engineering technology skills as follows:

- Product development and support
- Quality assurance
- Telecommunications and wireless systems development and support
- Biomedical instrumentation
- Test engineering
- Technical documentation
- Applications engineering
- Technical sales/marketing
- Biomedical computing
- Medical instrumentation
- Renewable energy.

These skills are applicable to positions wherein a theoretical knowledge of a machine, device, or component would not be essential for success. More on the "technical spectrum" later in Chapter 27 of this book.

Song: "Mr. Roboto"
As popularized by: Styx
1983

Graduate School

Studies have shown that engineers with advanced degrees will earn significantly more over the duration of their career than those without advanced degrees.

In 2013, the U.S. Bureau of Labor Statistics determined that those with advanced degrees earned approximately 27% higher starting salaries than those with undergraduate degrees. Therefore, it is important to consider the financial advantage of the advanced degree when planning the first part of your career to take full advantage. Typical masters degree programs are campus based and require approximately 30 credit hours of study. If a thesis is required, a maximum of six credit hours are given for thesis research. The remaining 24 credit hours are composed of course work. Some colleges offer a combination of BS/MS degree in a 5-year package.

For acceptance into their program, an application with fee, transcripts, letters of recommendation, and proof of English proficiency must be submitted to evaluate your eligibility. These requirements have not changed significantly but I have seen the cost per credit hour rise from $100 to approximately $1000. Now if you are over with this sticker shock, here are some possible economical alternatives.

Night school—This approach allows the student to be physically present in a classroom environment during the evenings. Course work is spread over a longer period of time to balance academics with a day work schedule.

Online programs—When being physically present in a classroom environment would be inconvenient, the course work can be completed at home thus avoiding excessive travel and time.

Graduate teaching assistantship—While working on your advanced degree, assuming you are of high academic standing, you will be teaching courses under the supervision of a professor in your field of study for about 20h a week. You may receive a "stipend" of up to $2000 per month.

Fellowship—While you obtain an advanced degree, you will be paid to conduct research in your field of study. Similar to a teaching assistantship, a high standing in academics is typically required for eligibility.

Today, due to the high unemployment rate, many students are trying to survive by staying on campus after receiving their undergraduate degree and seeking part-time work to help finance an advanced degree.

Song: "Almost Grown"
As popularized by: Chuck Berry
1959

Technical Career Survival Handbook. http://dx.doi.org/10.1016/B978-0-12-809372-6.00008-6

Day Versus Night School

Night school was the route I took following graduation when I accepted my first job at what is now Hamilton Sundstrand, Div. of United Technologies Corp. When it was time to register, I simply took a letter provided by UTC to Rensselaer Polytechnic Institute graduate center in Harford Connecticut and registered for my courses. I spread my course work over 4 years while maintaining my 40 h per week position as a design engineer. This educational benefit may be difficult to land in today's job market.

In reality, it is generally not critical to rush into a graduate program unless you want to remain in academia. It is the combination of experience and advanced degrees that will pay off in the long-term working in industry. I realized that obtaining funding for an advance degree through your employer may not be as common today as it was years ago due to the high cost. It was actually a deciding factor in making the decision on my first employer as I was completing my BS degree program. I recognized the value of this benefit allowing me to receive a full-time salary and obtain an MS degree paid for by my employer.

Unlike MS degree programs, Doctor of Philosophy (PhD) programs focus on basic research that expands the knowledge base of the field. The PhD program is designed to prepare a student to become a scholar and discover, integrate, and apply knowledge, as well as to communicate and disseminate it. These skills may lead to careers in social, governmental, educational, biomedical, business, and industrial organizations as well as in university and college teaching, research, and administration.

The PhD program emphasizes residency at the college or university for typically a two semesters of full-time enrollment although some schools will consider proposals for alternative residency. Both graded course credits and doctoral research credits are required plus a dissertation to demonstrate the candidate's ability to investigate phenomena in their field of study.

My experience working with PhDs in industry has been limited. Although not direct experience, while working at my first job with UTC, I learned that the Rensselaer Polytechnic Institute Hartford campus faculty was largely comprised of moonlighting PhDs that were employed during the day by several of the UTC divisions located in Connecticut. They were an unusual example of PhDs working in both the academic and industrial worlds.

While I was with Adtechs, a consulting engineering company specializing in hazardous and radioactive waste handling systems, we had Dr. Philip Baldwin, a chemical engineer, on staff to evaluate and recommend treatment methods depending on the chemical analysis of the waste constituents. In the manufacturing world, the demand for PhDs is fairly limited whereas the demand is higher in educational institutions or service agencies. In some instances, possessing a PhD might even limit

Technical Career Survival Handbook. http://dx.doi.org/10.1016/B978-0-12-809372-6.00009-8

employment opportunities in industry based on being "too heavy," meaning essentially overqualified.

As I mentioned in the previous section, there are obvious economic advantages in obtaining a degree at night while maintaining a job during the day. Of course, this requires an ability to balance of academics, work, and social life. However, there is another less obvious advantage that is often overlooked, I call "resume building."

While I used to whine about working summer jobs in an engineering environment as some of my buddies washed dishes at the beach, what really occurred was that I was unknowingly building a resume. Additionally, during my summer job as an intern at MikroPul, in Summit N.J. I was asked to stay on in the Fall to assist with some pressing R&D work associated with a patent law suit. I worked during the day assisting world renown consultant Dr. Frank Kreith, author of *The Principles of Heat Transfer*. That allowed me to take engineering courses at night in a nearby engineering college and save tuition money for my return to Virginia Tech to complete my senior year.

While the day job/night school combination set me back a year, not only did I generate some serious tuition money, I gained over a year's worth of technical experience plus character references for my resume. Also I believe it gelled my understanding of why I was in college and what I could expect upon graduation when I officially started my career.

When companies review resumes for entry level technical positions, they have to consider not only academics and job experience but they also look at how your past personal life might indirectly be beneficial to the job and the company. These might include community service, mission work, military service, and athletic competition or perhaps travel experience. But if you can show that you were able to handle both a school and work schedule, you will have one up on those that were strictly day students.

Of course, there may not be night courses offered by a recognized college convenient to your place of employment. That may be a serious obstacle preventing you from pursuing the night school approach. Best to check this out prior to accepting an employment position. In the next two chapters of this book, I will present two alternatives to night school, i.e., cooperative programs and internships.

Song: "Hard Day's Night"
As popularized by: The Beatles
1964

Co-Op Programs

10

A cooperative (co-op) student alternates time periods between on-campus academic life and on the job in industry. The rotation usually runs for one semester in industry on the job and then one semester in classes on-campus repeatedly and usually includes summer semesters. Many other variations are possible. Naturally, this rotation results in extending the overall time to obtain a BS degree for a typical 4-year program to 5 or more years.

For co-op program eligibility, many colleges like Pennsylvania State University (PSU) require completion of their first academic year with satisfactory results and complete entrance-to-major requirements. This is an ideal situation for the student that has very limited funding for college because they will receive compensation while working in industry. Compensation will then help finance the following semester on-campus and so on.

In most instances, the student will remain with the same industrial employer throughout the entire co-op program experience assuming there is sufficient work to support the student's employment. Some students may find a more desirable company at some point in the co-op program and therefore change employers. Geography may be a big factor.

Work assignments will be on a very fundamental engineering level for obvious reasons but nonetheless, the student will be expected to make a significant contribution to the company's projects, report on time, meet schedules, milestones, and deadlines. While the student must work out the specific hours on the job with their employer, most jobs start at 8:00 or 9:00 a.m. and finish at 4:00 or 5:00 p.m. A recent PSU survey of respondents indicated that average hourly salaries were approximately $19/hour depending on the curricula. This is the major upside to the program.

The major downside to co-op programs is the disruptive aspect particularly on college life. This pertains to participation in sports, fraternities, and extracurricula activities. Because the co-op student's schedule is based on alternating academia and industry, it will be difficult to be engaged in most campus activities. I have seen this first hand but priorities must be recognized and dealt with accordingly. However, generally these students are of the caliber that can deal with the disruption thus adapting and surviving. It will certainly make for an interesting choice for class reunion attendance.

Song: "Dream On"
As popularized by: Aerosmith
1972

Technical Career Survival Handbook. http://dx.doi.org/10.1016/B978-0-12-809372-6.00010-4

Internships

If you had asked me 30 years ago where to find an intern I would have told you to check the local hospital. Today, internships are not confined to the medical field. They provide the student with practical work experience in (or related to) their majors and provide a preview of things to come. Internships are usually a short-term work experience, i.e., 12–15 weeks in the Fall or Spring and perhaps a bit shorter in the summer, which is the most popular time. Sometimes more than one employment period/semester is completed with the same or different employer. This may occur during the sophomore, junior or senior years of college enrollment.

On the plus side, internships are often credited with completing for up to six academic credits. Interns may or may not receive compensation and less likely than for a co-op position (Table 11.1). However, it is important to note that internships are very dependent on the employer's needs and financial situation. From the employer's perspective, they have a low risk opportunity to observe the performance of a potential full-time employee on the job. However, they must be careful to provide work for the intern that is inspiring and represents a realistic preview of what lies ahead for the student should he or she eventually be hired by the employer.

Regarding compensation, there is an ongoing controversy over whether interns are actually "employees" who must be paid for all work time. In an attempt to clarify the question, the Department of Labor (DOL) has used a six-part test that includes the following:

1. Is the training similar to training given in an academic environment?
2. Does the internship experience benefit the intern?
3. Does the company derive immediate benefit from the intern's activities?
4. Does the internship displace regular employees?
5. Is the intern immediately employed at the conclusion of the internship?
6. Does the intern understand that he/she will not be paid for the internship?

It is recommended that you check if a potential employer's internship policy is in compliance with the minimum wages laws applicable at the time of their offer.

Work schedules for the internships of course are dependent upon the employer's needs and norms. Typically, they would require the intern to report for work between 7:00 and 8:00 a.m. and finish between 5:00 and 6:00 p.m. Check to be aware of any company dress code requirements. Other than free coffee, company benefits are not usually provided.

Jared Gollob, a young engineer with Industrial Turnaround Corporation (ITAC), interned for a manufacturing company while he was working on his degree at Old Dominion University in Norfolk Virginia. He admitted his internship consisted of

Technical Career Survival Handbook. http://dx.doi.org/10.1016/B978-0-12-809372-6.00011-6

Table 11.1 Intern Versus Co-op

Factor	Intern	Co-op
Regularly scheduled	Infrequent	Alternate semesters
Compensation	Possibly	Yes
Academic credit	Minimal	Yes
Benefits	No	Possibly

routine factory assembly work and did not receive compensation but received academic credits. Ultimately, he did not accept a position with the company after receiving his degree. However, according to the National Association of Colleges and Employers, nearly two-thirds of the interns receive offers from their companies upon graduation.

What students are eligible for internships? Typically, those undergraduates and graduates engineering students that complete at least one academic year and are in good standing. While the college may facilitate the placement of interns, that is not to stop the student from making arrangements with employers on their own. During poor economic conditions, internships may be difficult to arrange. Be certain to check with your specific engineering department for credits that will be allowed for the internship. Also, talk to an adviser on how your internship might affect any financial aid package that you may currently have.

Caution: If you interrupt your course schedule for an internship, be certain you do not miss an opportunity to enroll in an essential infrequently offered course that may be "required" for graduation.

Song: "Money for Nothing"
As popularized by: Dire Straits
1985

Alternative College Funding

Chapter Outline

As I stated previously, the cost of obtaining a degree has skyrocketed and there has been much angst over the cause and what to do about it. Compounding the problem has been the trend toward longer degree completion times. Student loans are not a solution but simply exacerbate the problem financially. According to a 2104 report by Complete College of America, only 36% of full-time students graduate on time. Some colleges are offering incentives and rebates for graduating on time. Although one might conclude that colleges gain financially by prolonged enrollment and trends like dual majors.

A study by education lender Sallie Mae determined that 74% of students work mainly in food service and retail throughout their college years. Surprisingly, middle-class students were more likely to work year-round than low-income students according to the study. As I stressed previously, working during college is a resume enhancer.

Other than paying cash, student loans or working your way through school, what are the other options? The simplest is some combination of these three, which generally are not dependent on academic standing. While obtaining an engineering degree might be tenuous, fortunately the expected stating salary will be one of the highest college majors. Recently, a student loan forgiveness program has been made available by the US government.

There are alternatives however, such as scholarships, grants, and GI Bills, all of which provide free money with no requirement for repayment. *Scholarships* are usually based on academic standing or merit and *grants* are often need based. However, often the two terms are used interchangeably. Both can come from federal, state, local governments, college, and private or nonprofit organizations. Usually, the student must first submit a statement as to the expected family contribution to determine the financial need.

Grants

The Federal Pell Grant is a funding method that is usually awarded to undergraduate students who have not already earned a bachelors or a professional degree. How much you will receive depends on your financial need, cost of attendance, and whether full

Technical Career Survival Handbook. http://dx.doi.org/10.1016/B978-0-12-809372-6.00012-8

or part-time enrollment is desired. The amount may cover up to a maximum of 12 semesters. The maximum Federal Pell Grant award was $5645 in 2013. Most Pell Grant money goes to students with family incomes under $20,000. A Federal Supplemental Educational Opportunity Grant (FSEOG) is available to students with exceptional financial needs. A TEACH Grant is available if you plan to become a teacher in a high-need, low-income area. If your parent or guardian has died in military service in Iran or Afghanistan, you could be eligible for a service grant. State and local grants are also available and are along the same lines as federal programs.

Scholarships

There are thousands of scholarships available from a wide range of private organizations, churches, employers, and institutions all with their unique criteria. Some of the criteria used in the selection process might include academic standing, athletic ability, musical ability, ethnicity, women, military service, financial need, and, in some instances, essay results. But be advised that the competition for scholarships is fierce. Many organizations only award a small number each year.

Let us look at the American Society of Mechanical Engineers (ASME) organization's requirements for scholarships as an example.

- Scholarships may be awarded to undergraduate and graduate students to enable or assist in their pursuit of a mechanical engineering or mechanical technology program.
- Scholarships are awarded annually to eligible ASME student members.
- Scholarship payments are paid directly to the student's college for the following academic year to assist with upcoming studies.
- To be eligible, students must be enrolled full time in mechanical engineering or technology studies that are accredited by the Accreditation Board for Engineering and Technology (ABET) organization.
- Students must have established their grade point average in their degree program.
- Scholarships have no citizenship or geographic requirements.

GI Bill

Beginning in 1944, congress appropriated funding for returning veterans to assist them with housing, living expenses, medical needs, education, and training. Post 9/11, veterans who served on active duty for at least 90 days may be eligible for full payment of their tuition and fees. They need not to exceed the maximum in-state undergraduate tuition and fees at a public institution in the state in which the student is enrolled. Tuition payments may be made by the Department of Veterans Affairs directly to the school on behalf of the student. Further, books and stipends of up to $1000 per year may be paid, which is prorated based on the student's payment schedule.

Eligible service members may transfer up to the total months of unused Post 9/11 GI Bill benefits or the entire 36 months of benefits if they are unused to family members. Check with the VA as the rules vary somewhat depending on the spouse, child, length of service, and active duty status.

Company Benefit

Many companies offer employee tuition reimbursement for those who work at least 20 or 30h a week and have at least 6 months continuous employment at the company. Reimbursement is usually contingent upon satisfactory completion of approved course work toward an associate's, bachelor's degree or even a master's degree program. This is a win–win for both the employee and the employer. The company encourages employment longevity, creates a more skilled workforce, and provides a hiring benefit. The employee gains credentials, receives payment for courses, and has the possibility of being promoted in the future upon completion of the degree program (see Chapter 9).

Song: "I Fought the Law and the Law Won"
As popularized by: Bobby Fuller Four
1966

Finding That First Job

![13]

Chapter Outline

This section is aimed specifically at finding a position immediately upon graduation. In Part III of the book, I will advise how to conduct a job search after you have gained experience and possibly additional education and training.

Hopefully, by the time you graduate, you will have some idea of which field you will seek for employment. Fields may include education, manufacturing, utilities, research and development, high tech, or consulting. This may be a result of summer, internship, or co-op employment. Or, perhaps you have a friend, relative or parent that has been influential. Finally, there may be a trend in industry that has resulted in a demand for technical degree graduates that seems appealing. That was my influence during the 1960s as the aircraft and aerospace industries were booming and seeking graduates with various engineering degrees.

If you are not certain about choosing a field of concentration, you may have to depend on another criterion to focus your initial job search. I will suggest the following but you determine the order of importance:

- Gaining experience in your chosen or preferred field

This can also be considered as "resume building," more on that subject later. While you have your sights set on a particular company in your chosen field, you may have to land elsewhere and gain some basic experience with a lesser company; in other words, "pay your dues." As you build your resume, your value to industry will grow hence so will your desirability to potential employers.

- Furthering your education

It may be very tempting to postpone pursuit of an advanced degree upon obtaining an undergraduate degree but here is the reality. The longer you wait, the more difficult it will be to get back into the study regiment for reasons such as marriage, children, hobbies, and other priorities. So if you deem it important for your future, start as soon as possible upon graduation. In fact, be certain that you will have the opportunity to go either full time or part time in the evenings with the company that hires you. For if you select a company located in a remote area, continuing your education may not be practical. Also, check to be certain that the company encourages furtherance of your studies, perhaps to the extent they are willing to pay all or a portion of your tuition upon satisfactory completion of the course(s).

• Seeking a good mentor/boss

An obvious example of this occurs in college when you find a particular professor that exemplifies the qualities that you find desirable in a mentor. This may lead to a graduate assistant position where you study under this scholar. Or perhaps, a supervisor during a co-op or internship program stood out as someone who could further your career.

During interviews you may have perspective employers, pay particular attention to the academic background of your potential supervisor.

Has he/she been shown to seek advanced studies?
Has he/she made reasonable advancements within the company?
Does he/she show a significant interest in you during the interview by asking key questions?
Does he/she have recent graduates under their scope of supervision?
Are those within his/her group pursing advanced degrees?

• Paying down student loans

According to the State Council of Higher Education for Virginia, the percentage of students with debt at graduation increased to 57% in 2011–2012. The median debt for that period was $24,354. Many student loans today have long-term payback periods, almost equivalent to home mortgages of a few decades ago. Although, monthly payments may be minor compared to starting salaries, you may not be comfortable having the debt hanging over you. This may be an important consideration causing you to move in with your parents or relatives and finding a job within a reasonable commuting distance. In this way, you can minimize your living expenses to accelerate your loan payoff.

Where to Look for Your First Job?

The Obvious

Consider finding your first job, your first full-time job. Seek employment with your co-op, internship, or summer job employer and let them know about a year in advance of your graduation that you are highly interested. Many companies make provisions in their budgets and headcounts well in advance. Check with your campus placement office to find out what companies may be coming to your campus to interview candidates for degrees. If the campus interview goes well, usually a company visit is the next step. You may not necessarily meet the actual hiring authority until you visit the company site. More on that subject later.

The Not So Obvious

Network through a friend, family member, relative, or previous graduate—expect that they can provide you with the name and phone number of the company and the hiring authority. Explain when you call the company, you are requesting an opportunity to visit the employer for possible employment.

Frank Brooks was one of my coworkers at General Electric Gas Turbine Department during a time period when entry level jobs for recent grades were scare. With his recent electrical engineering degree from RPI in hand, Frank interviewed GE human resources and awaited a response. Meanwhile, Frank was introduced to his father's manager at GE. Following an informal lunch interview, he offered him a job as a piping design engineer. Surprisingly, 2 weeks after he began employment, he received a letter from the GE HR department advising they were not going to make him an offer. In large companies, communication between the HR and other departments can be a less than optimum. In spite of the rocky start, Frank received several promotions and ultimately become manager of applications engineering until his recent retirement from GE.

Online or newspaper wants ads and job fairs—these tend to be localized so you may have to visit areas or search for the information online to determine who is hiring and where. Tommy Atran, after graduating from Virginia Military Institute, simply returned to his home in Richmond where he attended a nearby job fair. There he learned about ITAC, an engineering consulting company that was seeking mechanical engineers. Subsequent to an on the spot interview, he visited the company and received an offer for employment, all within a few weeks.

Check with employment agencies—They place candidates with employers for a percentage of your hourly rate. Seek out those agencies that specialize in placing technical personnel. This is likely to be a temporary employment opportunity but may lead to eventually being hired full time by the employer at which time the employer will probably pay a finder's fee to the agency. All the while, you are building a resume.

Technical recruiters—They place candidates in full-time, long-term positions also for a fee but generally prefer to work with experienced technical personnel or those with a unique set of credentials. More on "headhunters" in Part III of the book.

Song: "Lodi"
As popularized by: Creedence Clearwater Revival
1969

Geographic Factors

14

Chapter Outline

Where Are the Jobs?

In the past, I have always been a bit envious of the civil and electrical engineers I have worked with because it seems that no matter where they chose to live, they can find companies that design and build roads, bridges, or distribute electrical power. And in recent years, you could add computer technicians to the list as currently, they are in demand everywhere and not so with chemical engineers, nuclear engineers, metallurgical engineers, and many other disciplines. So where do they look for employment?

Technological employment opportunities ebb and flow with the economy. When I was graduated in the 1960s, the job market was extremely hot. Students with average grades could easily find positions with top companies. That was then, but now it is vastly different. If we assume that technology jobs will exist in areas where the most favorable business climates are, then perhaps one would examine those areas when seeking employment. So what are the factors that result in a favorable business climate? Here is what *Site Selection*, a magazine of corporate real estate strategy highlighted in a November 2011 article as the key factors in their survey:

- Right-to-work state
- No state income tax
- Few: regulations, state fights with OSHA and EPA
- Government makes it easy to do business
- Readily available workforce and existing facilities

Based on these criteria and others, Site Selection determined that the top states were ranked as follows:

1. Texas
2. Georgia
3. North Carolina
4. Virginia
5. South Carolina
6. Indiana

Technical Career Survival Handbook. http://dx.doi.org/10.1016/B978-0-12-809372-6.00014-1

7. Louisiana
8. Tennessee
9. Ohio
10. Florida

So would not it make sense to explore these states when beginning your job search? But before you hop on a plane bound for Texas, read on.

Where Are Manufacturing Centers?

Here is another way to look at technological employment opportunities. According to March 14, 2012 report in 24/7 Wall Street. 10 US cities/regions top the list where manufacturing is booming. They include the following:

1. Milwaukee–Waukesha–West Allis, Wis.
2. Youngstown–Warren–Boardman, Ohio-Pa.
3. Toledo, Ohio
4. Greenville–Mauldin–Easley, S.C.
5. Grand Rapids–Wyoming, Mich.
6. Louisville-Jefferson County, Ky-Ind.
7. Columbia, S.C.
8. Ogden–Clearfield, Utah
9. Cleveland–Elyria–Mentor, Ohio
10. Charleston–North Charleston–Summerville, S.C.

Later in Part II, I will look at "technology centers" focusing on regions where similar industries are located in yet another way to survive a job search.

Song: "Ramblin' Man"
As popularized by: Allman Brothers
1973

Interviews

Chapter Outline

The candidate experience is a series of steps that a company takes to hire a recent gradu-ate and can be intimidating for the candidate. Employers have many more tools available today in the digital age than in the past. But inevitably an interview with the candidate will occur. Several venues are likely for the interview process, here are a few:

Campus

Many engineering colleges offer students an opportunity to interview prospective employers on campus for internships, co-op, and permanent positions. These inter-views are typically scheduled in advance and conducted during a specific time period within each semester as determined by the Office of Student Affairs. Companies will post their on-campus dates in advance. They may require that student resumes be submitted to employers in advance online. Interviews are conducted one-on-one, employer to student at the student affairs office.

Telephone

Often a telephone interview is conducted one-on-one between the employer hiring authority to student following the employer's thorough review of the applicants resume. This is a convenient, low cost way of determining if the student is a good fit for the posi-tion. However, it is usually conducted with a specific agenda similar to a face-to-face interview re: Q&A. Although I have not experienced one, a Skype or video interview might be requested by the potential employer as an alternative to the telephone. If so, dress accordingly, keep looking at the camera and pick a back drop that is clutter free. Potential employers take phone and video interviews quite seriously and so should you!

Technical Career Survival Handbook. http://dx.doi.org/10.1016/B978-0-12-809372-6.00015-3

Neutral Location Interview

While this may be uncommon, I have experienced an interview conducted at an airport. The recruiter and I happened to be traveling and a particular airport was a convenient meeting place. The interview ultimately led to a position with the company following an employer's facility interview.

Employer's Facility

This is the ultimate location for the candidate to grasp a full understanding of the employment opportunity and is usually where a successful telephone or campus interview leads. It will typically include the following:

- Meeting with human resource personnel to discuss benefits—Let the employer arrange for this, do not request a review of benefits unless an offer is eminent.
- Introductions to future peers—This is a way to make you feel comfortable and get a head start on team building among peers.
- Meeting with the hiring authority—This is the decision-maker who may or may not be your immediate future supervisor. You can ask if there is a written job descriptions if they do not bring it up first. It should address how the position can aid the candidate's progression in the organization.
- Plant/facility tour—Try to determine if this is an environment that you find interesting and raises your curiosity.

Panel

If this is the case, you will meet with three to five people such as the HR representative, your supervisor, manager, peer, and others that you might potentially interface with. In this scenario, all participants hear the same thing versus one-on-one interviews. As for you, the interviewee, you will see how they interact with one another and you will gain insight into the company culture. Make sure you write down all the names of the participants and their titles going around the table so you can address them accordingly and contact them later if desired.

Do not be totally shocked if you are asked to take a brief aptitude test. This was my experience when interviewing General Electric Jet Engine Division in Ohio. What I remember most was taking the test even though I had already concluded that I was not interested in the position. Take it, it will not hurt you.

Here are what I consider my top 10 important tips regarding the interview process:

1. Turn off your cell phone.
2. Dress for success.
3. Bring examples (portfolio) of your past projects and be prepared to discuss them briefly.
4. Take notes.
5. Research the company and have pre-prepared questions.

6. Have an advanced idea of how to express your strengths and weaknesses if you are asked.
7. Practice being interviewed and answering anticipated questions concisely and briefly. Follow the example of a news anchor.
8. Avoid discussing your salary requirements, say your "open." Taboo topics such as age, religion, whether you drink, or smoke should be avoided.
9. Express your interest and ask to be considered for the job at the conclusion of the interview.
10. Follow up with a letter/email expressing your interest in the job.

It does not hurt to sign up for an interview even with a company you may not favor, do it for practice. Not only is it good experience but also it will allow you to be a little more comfortable when the right job interview comes along.

Song: "We Are the Champions"
As popularized by: Queen
1977

Opportunities for Women

16

This subject has always puzzled me and therefore I chose to include it in this book. When I was a high school student, I knew many girls that excelled in math and science and therefore it was surprising to me that so few elected to pursue technical careers. After all, technical careers generally do not require brute strength, getting muddy, or handling explosives. But the sum total of my work experience has resulted in knowing only 10 women in engineering or design positions. I think I am now beginning to understand why.

1. Graham Allan, professor in chemical engineering at University of Washington wrote in Consulting-Specifying Engineer magazine that it is well established that only one-tenth of employed engineers are women. I would have guessed less. He further writes that women should be attracted to the vocation because of their unique creative ability and not be discouraged by lower performance in science and mathematics when compared with their male counterparts.
2. A 2011 research study by the American Society of Engineering Education (ASEE) found that the percentage of female engineering graduates accounted for approximately 18.4% of the BS degrees awarded in total versus 18.1% in 2010 (Fig. 16.1).

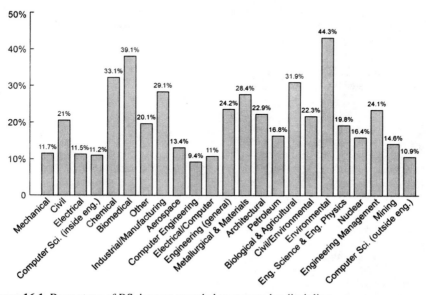

Figure 16.1 Percentage of BS degrees awarded to women by discipline.
Source: American Society of Engineering Education data 2011.

Technical Career Survival Handbook. http://dx.doi.org/10.1016/B978-0-12-809372-6.00016-5

3. Often mathematics and science are grouped together in what is referred to as the acronym "STEM" (science, technology, engineering, and mathematics). It is important to note that a 2011 US. Census Bureau publication titled *Disparities in STEM Employment* concludes that nearly 1 in 5 female science and engineering graduates leave the technical labor force versus only 1 in 10 males. Many receiving BS in engineering degrees pursue nonengineering careers such as law, education, health care, and even accounting. I knew three male engineering friends early in my career who left technical positions to become lawyers.

4. According to the Census Bureau report, mechanical engineers have the lowest female representation at 6% women.

5. In a recent study compiled at Harvey Mudd College in California by its president, the number one reason why young women do not chose technical careers is that they think they are not interesting. The second reason, the young women do not think they would do well.

Back to my dilemma, since my work has focused on mechanical engineering, the study would conclude that for every nine male mechanical engineering degrees earned, there would be one female degree earned. Also, more mechanical engineering degrees are awarded than any other discipline. So why have I known so few women engineers? However, I also recognize that there has been more press in recent decades highlighting chemical, biomedical, and "save the environment" issues. Perhaps the front page stories constantly focusing on these problems have motivated women to pursue these fields of study thereby explaining the survey results.

One of the consequences of this scarcity of women in technical field has created a great opportunity and a high demand for their recruitment particularly companies advocating diversity in the workplace. This often has companies asking recruiters to favor women in their employee searches. Particularly, this is the case at the CEO level where only about 4% of CEOs of Fortune 100 companies are female. It would seem that in the future, improvement in promoting technical careers for women will be necessary.

Song: "She Works Hard for the Money"
As popularized by: Donna Summer
1983

Part II

Employment Challenges

Experience

17

Your first job may or may not be indicative of the focus of your career but it is definitely where you build your resume and add value to your abilities. Experience begins with your first job as defined by your discipline, industry, and specialty. Let us take them one at a time.

Discipline

This is your academic credential such as bachelors, masters, or PhD degrees generally pertaining to one area of study such as mechanical, electrical, or civil engineering. Dual majors are fairly common today in liberal arts degrees but uncommon in engineering as it may require an additional year or two of college study.

Industry

This is where you land your first job working in the automotive, aerospace, computer, electronics, or other fields. I would also include academia for teaching assistants in colleges or universities with the potential to become full professors.

Specialty

This is your focus within your discipline and the industry which you are employed. Examples include analysis, design, quality control, testing, safety, projects, sales and service, and the specific skills you will develop.

Technical Career Survival Handbook. http://dx.doi.org/10.1016/B978-0-12-809372-6.00017-7

Let us assume you have worked 5 years in your first job and you decide to change jobs and relocate to a particular region. You can summarize your first 5 years as BS electrical engineer in the aerospace industry specializing in avionics design. From strictly an experience standpoint, aerospace employers who manufacture avionics will greatly appreciate your background and perhaps pay generously to bring you onboard.

So what if you should decide to make a change in your specialization to electrical generators? You may encounter a financial setback as a result of the switch. This will likely not put you back to square one but you will certainly need to express flexibility and a strong desire to broaden your experience when you interview. You may be successful on landing your new job based on your 5 years of experience in the industry alone. With longevity and while maintaining your specialty, you will become valuable to that one company and also their competitors. But by changing industries, your specialty may make you less valuable. Here are examples of some career changes:

1. This is an example of an industry change based on a downturn in the aerospace industry I experienced. My first 4 years on engineering employment where summarized as BS mechanical engineer in the aerospace industry specializing in rotating machinery design. In 1970, with my MS degree earned at night, I successfully changed to the power industry (General Electric Gas Turbine Division) specializing in machinery design and received a respectable salary increase. Several years later, during a turndown in the power industry, I successfully transitioned to the pump industry specializing in machinery design management with a significant increase in salary and benefits.
2. This is an example of both an industry and specialization change. The most radical career switch I had was due to a layoff/downsizing while I was director of engineering at Sundstrand Fluid Handling (SFH). Because my family and I were focused on relocating to Virginia, I decided to transition to a technical recruiter aka "headhunter" and purchased a franchise accordingly which included training and support. That major reboot resulted in opening a furnished office, hiring recruiters and working on commission. After 1 year, I decided to return to the engineering field resulting in a less radical transition.
3. Here is an example of a successful industry transition. Recently, I met Joseph Moody, an electrical engineering graduate who also possesses an MBA from University of Michigan. He spent the first 20 years of his career in the automotive industry wherein his specialty was in manufacturing. Following his automotive experience, he switched to the petroleum industry for 9 years while maintaining his manufacturing specialization. Then recently he became president of a research and development center focused on advanced manufacturing technologies.

In summary, transitions in discipline, industry, and specialization are a common occurrence and are often necessary for survival. You may make some good decisions and some you will regret. You cannot possibly know all the details about a future position before deciding to jump in. However, in many instances, job changes result in building your resume accompanied by significant increase in salary and benefits provided the transition enhances your overall background.

Caution: Radical transitions such as changing from the aerospace to the automotive industry might result in a career reboot. Several of my old college friends made radical moves transitioning from engineering to law. One of them, Bill Wells, became a patent attorney wherein his BS mechanical engineering was a major resume enhancement.

Song: "American Pie"
As popularized by: Don McLean
1971

Work Direction

18

You will likely begin your first job reporting to an assigned supervisor. In many companies, the title "supervisor" describes a person who provides work direction for his/her subordinates. He/she will likely have considerable experience in the skills which you will be acquiring and will likely provide day-to-day guidance and training. Supervisors generally have training both in leadership and in technical skills. Many mid to large companies will provide leadership training for groups of similar supervisors and individual contributors in anticipation of being promoted to supervisory positions.

Many companies use the term "lead engineer" to designate a specific engineer who provides daily work direction on a project but does not have authority to evaluate individual's performance or salary level. This would be the responsibility of the supervisor but perhaps based somewhat on input received from the lead engineer. Often the lead engineer is of one discipline but provides work direction of different disciplines where in their work is somewhat secondary or supportive to the lead engineer's project.

For example, assume a lead engineer is responsible for a pumping station design, he or she prepares equipment specifications and selects the pumping equipment. He/she may provide work direction for a civil engineer to provide a building design, an electrical engineer to design electrical panels, a control engineer to design a control system and a computer-aided design operator/designer to model the building and equipment arrangement. All of these support people likely report to a supervisor or manager of their own discipline who is familiar with their specialty and therefore better suited to evaluate their performance.

Regarding hiring and salary adjustments, some companies delegate these responsibilities to supervisors where other companies assign this responsibility to managers. Some would argue in favor of the supervisor having this responsibility to provide enforcement ability and credibility so as to praise and/or criticize subordinates. However, the supervisor must necessarily have training in performance evaluation in addition to discipline skills thereby wearing two hats. Thus one would argue that performance evaluation should be assigned to managers who serve more in an administrative function.

My first supervisory position in industry was a hybrid and titled "product manager." I provided work direction on a daily basis and evaluated personnel performance formally on a semiannual basis. But I was also performing engineering work right alongside my subordinates. Salary adjustments, performance evaluations, and hiring decisions however were allocated to my immediate boss who was responsible for managing other product managers (see the next section for more on the manager's role).

Song: "It's Five O'Clock Somewhere"
As popularized by: Allan Jackson and Jimmy Buffett
2003

Technical Career Survival Handbook. http://dx.doi.org/10.1016/B978-0-12-809372-6.00018-9

Manager

19

In a classical sense, an engineering manager is responsible for hiring, firing, promoting, and evaluating the performance of his or her subordinates. Some of his or her subordinates may be titled "supervisor" and assist him with these responsibilities. Other duties may likely include some or all of the following:

1. Preparation of a 1- or 5-year business plan describing the work to be conducted with the appropriate goals and objectives.
2. Communicating progress toward goals and objectives to upper management.
3. Preparation of the yearly budget including salaries and expenses.
4. Preparing a manpower plan.
5. Establishing a training plan for self and subordinates.
6. Maintaining a knowledge of advancements in the industry and the state-of-the-art.
7. Representing the company in trade associations and conferences.
8. Approval of all capital and expense spending for the department.
9. Recognize and pursue patent and trademark opportunities.
10. Evaluate competitive products and recommend responsive action.
11. Prepare and maintain job descriptions for subordinates and future hires.
12. Make presentations on progress and projects to management, clients, and subordinates.
13. Mentor a backup and provide training.
14. Recognize accomplishments of outstanding employees.

It is not unusual for a manger to report to a manager particularly in a midsize to large company. Although my experience was the use of the title "director" to describe those in charge of multiple managers. Directors might report to a "vice president," "general manager," or even the "president."

While one might assume becoming a manager would be welcome by most technical personnel but oh contraire. When I was promoted to director of engineering while working at Sundstrand Fluid Handing, my first task was to select a manager to fill my previous position. There were several senior engineers who were eligible so searching outside the company was unnecessary but my internal choice took an interesting turn.

After reaching a decision, I selected Bill Mabe as a senior engineer, who was quite reluctant to accept the offer. I withheld the mention of salary increase from him although I am certain he realized that would it follow. Then I realized that perhaps he felt somewhat uncomfortable if he were to be elevated above his peers whom he had a long and satisfactory working relationship. Additionally, a manager's responsibilities are mostly administrative and many technical personnel are unwilling to make that transition from what they are comfortable. But after considerable encouragement on my part, he agreed to accept the promotion and with time, he proved to be competent in his new role.

Technical Career Survival Handbook. http://dx.doi.org/10.1016/B978-0-12-809372-6.00019-0

Often candidates for promotion receive advanced training to prepare them for the eventuality of moving up in the near future and enhancing their possibility of survival. This also serves as a means of signaling subordinates that their superiors believe they have the potential for advancement. Or, on the other hand, the candidate may express displeasure when given the opportunity to move in the management direction.

Song: "Teach Your Children"
As popularized by: Crosby, Stills and Nash
1970

Management Styles

Chapter Outline

Inevitability that you will encounter a wide variety of management styles during the course of your career. They will be evident in all levels of supervision from supervisor, manager, director et al. There will be little you can do to change their behavior but I will suggest some actions on your part to deal with the styles you may encounter. Here are a few that I have recognized in the past from most desirable to least desirable from a subordinate's perspective.

Coach

This supervisor/manager will provide you with both technical and behavior support. He will motivate you, deal with conflicts, and let you know that you are performing well. He will also help you resolve issues you might encounter with others, i.e., go to bat for you. Your best bet for a boss early in your career.

Micromanager

Sometimes referred to as a detail person. Every task you perform will be carefully scrutinized and evaluated versus known company procedures and policies. In some ways, he/she may be acting as your parent if you are early in your career. Probably best if you meet with him or her frequently and/or provide daily updates on your progress at least until he or she is comfortable with your performance.

Technical Career Survival Handbook. http://dx.doi.org/10.1016/B978-0-12-809372-6.00020-7

Conservative

This manager "goes by the book" and follows company policies so as not to upset his superiors. He likes paperwork, follows a routine but is somewhat insecure, and therefore avoids involvement. Be certain your tasks are clearly defined and schedules are well established.

Ambivalent

This manager yields to pressure easily and prefers standard approaches to problems. It is unlikely that he will go to bat for you in a crunch situation by taking a strong position. Best if you document your calculations well and have your references authenticated. He will want them as backup if questions arise.

Dodger

This manager is in the clouds, unaware of conflicts, and may be overly concerned about his retirement. He/she may cause you to fend for yourself leaving you without clear goals and objectives. Try to schedule daily, brief, impromptu meetings to force him/her to provide guidance and reassurance.

Dictator

Sometimes referred to as an authoritarian or a bully. He/she will have you working long hours and feel guilty when you take time off. It is best to communicate with him/her often, even though as uncomfortable as it maybe. Most importantly, document you work well. You may also want to prepare for your next career move.

In most cases, your future will be dependent on your ability to survive the supervision under the types of management listed above. Conclusion, learn to adapt.

Song: "Piano Man"
As popularized by: Billy Joel
1973

Positions—Open Versus Created

21

You may assume that companies desire to fill an "open" technical position so that they can complete a development program or add a new product or feature to their portfolio. In most cases that is the case. It is also a common occurrence that a position is open because of company growth, an employee quit, or the position is newly "created." If you are a possible candidate for any position, I recommend you question during your interview why the position is being filled as it may help you negotiate the offer particularly if you possess a unique skill that they are searching for.

Back to the question regarding the created position. If this is the case, there may be a technology deficiency in the organization, which must be filled to move forward. For example, while I was working for a Japanese-owned consulting engineering company Adtechs, that specialized in designing hazardous and radioactive waste treatment facilities, management recognized the complexity of the waste streams that we were trying to treat with various types of process equipment. Therefore, a position for a senior level chemist was created and later filled by Dr Philip Baldwin, who had a background in waste stream chemistry. This created position was unique to the company and the salary and job description were prepared after determining the experience and capabilities of the newly hired person. For consistency with company salary structure, it was determined that the newly hired chemist would serve as a staff person reporting to the vice president of engineering.

When a new position is created, a salary range must be established that is compatible with the job description and other existing positions within the company that are compatible. Obviously, the level of supervision must be determined as well. You may frequently see the term "open" referred to when a company posts a want ad and they may not have yet established a salary range. Perhaps they are seeing what the market has to offer.

As opposed to the created position, the "open" position is generally well established and often a result of a quit or promotion. And it follows that there is usually a clearly defined job description and salary structure associated with the position. More on those topics will be discussed later. In either case, a company should make an open or created position known to its employees generally by "posting" the position on the company website or bulletin board well in advance of looking outside the company. If employees are not made aware of the opportunity, major discontentment may follow.

It is not unusual for a company to go on a "fishing expedition" to fill an open position. This occurs when they think they have an internal candidate for the position but there may be some reluctance on the part of management to promote him or her. Reasons might include performance history, salary, the wrong specialty for the job, or other factors. The employer proceeds to look outside the company to see if the ideal candidate exists. Then they make comparisons internal versus external

Technical Career Survival Handbook. http://dx.doi.org/10.1016/B978-0-12-809372-6.00021-9

before reaching a conclusion. They may determine that external candidates are way too expensive and/or would result in excessive relocation expenses. What happens next? They go back and settle on the internal candidate and strike a deal. If you are an internal candidate and experience this process first hand, do not be shocked, be humble and survive.

Song: "Career Opportunities"
As popularized by: The Clash
1977

Specialties

22

Years ago, someone once told me that there is no such thing as a permanent position. Possible back in my father's day that was not the case however now it is. The current terminology for a permanent position is "employee" also known as a survivor. This is to make a distinction from a "contractor." More on that will be discussed in the next section.

The technical positions or specialties that exist within companies can be categorized and some generically defined as follows:

Research—Utilizing fundamental scientific concepts, these personnel provide calculation methods and formulate materials for potential product composition/construction.

Analysis—These are the number of crunchers who utilize various types of software to make performance predictions and/or provide design data for product design and evaluation.

Development—Using established engineering principles, performance predictions, and/or design data, these employees create new product prototypes and product features for potential markets.

Design—Based on newly developed product prototypes, design data, and commercial materials, products and product features are configured for future production and marketing.

Projects—These personnel procure prototypes, arrange for verification testing, evaluate results, and make recommendations for product and product feature modifications and production.

Manufacturing—Utilizing known manufacturing processes, they produce prototypes and production products in accordance with a predetermined schedule based on desired quantities.

Quality control—Using measurement standards, tests, and equipment, they ensure that products are produced to desired standards and tolerances.

Supervisor—Works closely with the subordinate to be certain goals and objectives are fulfilled. May administer salary and benefit adjustments, review performance, and hire additional personnel. They are usually of the same discipline as the subordinate.

Project or product management—Guides other engineers and technicians toward the completion of a project or product development program in accordance with an acceptable schedule and budget.

Testing—Using test equipment, standards, and instrumentation, engineers and technicians evaluate product and product feature performance compared with acceptance standards.

Application—Determine which product fits a known application and evaluates how the product will perform using known methods and parameters.

Service—Perform field tests and evaluate data to determine if the product is operating in a satisfactory manner or make modifications to correct deficiencies.

Trainer—Establish training materials, cutaways, and video to familiarize employees and users with the technical aspects of the products.

Sales—Relates product performance and features to the end user and predicts demand of products for manufacturing production scheduling.

Technical Career Survival Handbook. http://dx.doi.org/10.1016/B978-0-12-809372-6.00022-0

Your corporate survival will depend on your ability to land in the right specialty. So get a clear understanding of the expectations associated with the position.

Song: "Piano Man"
As popularized by: Billy Joel
1973

Contractor

23

Many companies meet their goals and objectives not by adding addition staff but by supplementing their workforce with "temporary personnel" as needed. The advantage to the employer is not committing salary and benefits on a long-term basis but at the same time satisfying the peak manpower demands. For example, chemical plants often plan shutdowns called "outages" while new equipment is being installed or debottle necking (increasing throughput) is taking place. Introducing new product lines into the market may also require a spike in employee activity. Let us look at how these temporary demands are met.

In a technical setting, a contractor is not a permanent employee but is hired to perform many of the duties of a company employee. Working hours will vary depending on the company's needs. Contractors are basically brought in on a temporary basis and are not eligible for the employer's benefit program. Here are a few characteristics of contractors:

1. They are typically highly experienced in the field and specialization they are hired to perform. They do not typically fill entrée level positions.
2. Often their experience is comprised of closely related work and in some instances that of a competitor.
3. They may have been laid off by a previous employer and have found it difficult to obtain another full-time employee position.
4. They enjoy the flexibility of working for a certain time period then taking time off to pursue other interests.
5. Because they tend to fill a position abruptly, they are flexible, generally low maintenance, and self-starting.
6. Most are not obsessed with long-term commitments and retirement.

Third parties or "agencies" provide hourly pay and sometimes offer benefits to their employees and place them in industry return for a fee paid by the employer. The difference between the employer fee and the contractor's employee hourly rate plus benefits is the agency's profit. Typically the benefits provided are minimal but hourly rates may be quite high. These agencies usually specialize in various technologies and often founded by former employees of the industries they serve. What if the company desires to hire a contractor for an employee position? While this happens frequently, there is likely a predefined contractual agreement that requires the company to pay the agency a "finder's fee." This is usually a percentage of the first year salary.

I was a contractor while working on a project for Dominion Virginia Power for a period of 10 months. The agency relied on my resume but I also was required to meet with the Dominion project manager Emil Avram who subsequently gave hiring approval to the agency. Having just finished employment with Day & Zimmerman

Technical Career Survival Handbook. http://dx.doi.org/10.1016/B978-0-12-809372-6.00023-2

consultants who lost their alliance contract with Honeywell, I was available to start immediately. The project was completed in about 1 year at which time Emil prepared an awesome letter of recommendation for me. This was very helpful in explaining my contribution to the project to perspective employers who might be leery of the short-term nature of the job.

Some refer to employees like I was as "hollywood contractors." Whereby a leader/manager assembles a team of talent to work on a single project for a set duration. Then when the work is completed, closes out the project and everyone moves on to the next gig. This is much like hollywood assembles a team of actors, actresses, wardrobe personnel, gaffers, cameramen, directors, and writers to produce a single movie. After it is completed, the movie is released and everyone moves on.

Companies often prefer using agencies rather than paying an individual directly from an administrative perspective. By using an agency, the company simply cuts a single check for the pay period and they do not have to be concerned about deductions for the individual's FICA, Medicare, and various taxes. Similarly, the company may utilize one agency exclusively for all contract positions further simplifying their accounting.

If the company choses to avoid an agency and hires a "free agent" directly, the internal revenue has established extensive criteria to determine whether an employee is truly a free agent or a full-time employee. The difference has significant tax ramifications.

In recent years, there have been rare cases where contractors have been working for a company for many years and decided to claim they were therefore entitled to certain benefits, i.e., the company pension plan. A takeoff on the common law wife principle. To counter that potential problem, companies like my former employer Dominion Resources in Richmond will limit the contractor's period of employment, release them temporarily, and then bring them back onboard based on a new contractual agreement/period. I have known contractors who were able to survive this employment situation for many years.

Song: "Opportunities"
As popularized by: The Pet Shop Boys
1985

Consultant/Self-employed

24

Self-employed technologists or consultants are usually paid by the hour, do not receive benefits, and can be terminated at will by the company. While I was director of engineering for SFH, we hired a well-known consultant and author in the rotating machinery field, the late Dr O.E. Balje who truly was a "rocket scientist." We had a specific product that we were developing and required outside expertise in aerodynamics for the short term. In addition to paying him an hourly rate and purchasing his computer design software, we contractually agreed to utilize his services for a minimum number of hours each year (Fig. 24.1). This allowed him to plan his support for other clients as well.

How does one become a consultant? Like in many cases, some by accident and some by design. But the real answer is when one realizes they can make a reasonable living from doing so. In the case of Dr Balje, he wrote a book titled *Turbomachines, A Guide to Design and Theory*. He covered the subjects with principal emphasis on machinery selection, preliminary design layout, and background theory as a seasoned expert in the field. As a result of authoring this book, Dr Balje became quite famous but he had to make it known that he was available for hire, hence he had to market his services.

Dr Balje had a specific knowledge that was very marketable. If you are considering becoming a private consultant, you will need to determine if your specialties are in demand. Are companies willing to go outside their organization to enlist your skills? Do you know how to contact those who would hire you and are you willing to go out and actively market yourself?

Figure 24.1 A consultant's computer software was utilized to design this rotor for 60,000 rpm compressor operation.

Technical Career Survival Handbook. http://dx.doi.org/10.1016/B978-0-12-809372-6.00024-4

Many technical people quit or retire then become consultants to their former employer. This works well provided that no bridges are burned on their way to retirement. As a consultant, you will also need to be cautious supporting other clients to avoid passing on competitive information. My past employer Dominion Resources would hire back their retired employees but restrict the employment period to 1000 h per year. The previously listed six characteristics of contractors also apply to consultants. However, they are usually not signed with an agency and therefore are billed as an independent "vendor."

Note a word of caution when enlisting a consultant's services. Be aware that information you *provide* to a consultant might possibly be passed on to another client. Or, vice versa, the consultant may inadvertently be passing on proprietary information pertaining to your competitors. Therefore, be certain to consult with a corporate attorney as to how the scope of the consultant's services will be defined.

Finally, should you choose to become a consultant, you will therefore be self-employed. You will need to manage your cash flow, budget your time, market your services, obtain legal advice, and provide your own benefits and liability insurance to survive. Most importantly, you will need to prepare a business plan. Failing to plan is planning to fail.

Consulting engineering *companies* will be discussed later.

Song: "Taking Care of Business"
As popularized by: Bachman-Turner Overdrive
1973

Part Time

25

A part-time employee is an employee of a company who typically works less than 40 h per week and may have certain company benefits. Their work can be specific in days and times or vary based on the needs of the company. Currently the news media sites companies that are hiring people to work less than 30 h per week part time to avoid paying for the employee's health insurance. There is not much evidence that this pertains to technical personnel currently.

Many technical people quit or retire then become part-time employees for their former employer.

Thus the advantage to the company is having an employee who is familiar with the company products and services right out of the gate but since the employee is part time, payroll expenses are obviously reduced.

In an inverse way, both agencies and consulting engineering companies may wish to hire technical people who quit or retire from a company as well. They can then offer the services of these individuals to the companies from which they left as they may be very familiar with the company personnel, policies, projects, documents, and work load.

From an individual's perspective, here are some of the pros and cons of part-time work:

Cons

- reduced income and benefits by working fewer hours
- lower ranked seniority than full-time employees
- more likely to be assigned less desirable projects/tasks
- less likely to be promoted or receive regularly scheduled salary increases
- may feel a bit removed from the team
- the assigned work load may fluctuate significantly

Pros

- more time to pursue a second business
- more time for leisure activities
- more time with family
- enroll in college classes or seek special training requirements
- receive income rather than unemployment checks.

Caution: As a part-time employee, be aware that the irregularity of position is a risk for your survival.

Song: "Shiftwork"
As popularized by: Kenny Chesney
2007

Technical Career Survival Handbook. http://dx.doi.org/10.1016/B978-0-12-809372-6.00025-6

Temporary

26

Some employment companies are known as "agencies" or "workforce augmentation specialists" or "staffing companies" who place temporary technical personnel. These placements are typically for a few months and generally are for technical but non-engineering positions. Benefits are occasionally offered by the staffing companies but are usually minimal in scope. Work hours are generally 40 h per week minimum and overtime may be required. Often temporary employees simply fill in for employees on leave or in remote assignments. The hourly rate may be higher than the full-time employee but relocations expenses will not likely be covered.

Here is an example of a recent ad:

"Ajax Corp. has an immediate need for TEMPORARY Engineering Technician for approximately a 3–6 month assignment to work in our Technical Center within Research and Development. This temporary position will be a contract position through a contract staffing company."

Note the ad refers the potential candidate to go through a staffing company. Therefore, anyone pursuing this position should be prepared to apply through the particular staffing company preferred by the corporation who will likely divulge their name of the employer when you apply for the position. The staffing company may also advertise for the position without mentioning their client company's name aka "blind ad."

Besides working through an agency, a second type of temporary position is working directly for the employer/company. In that way a temporary employee is covered by the company's liability and other insurance policies, including workers compensation. In this way, the company may save an agency fee but must be prepared to search and interview temporary employees plus deal with the associated payroll requirements versus a single payment to the agency.

It goes without saying that temporary positions are risky but they may be essential for your long-term career survival.

Song: "Ventura Highway"
As popularized by: America
1972

Technical Career Survival Handbook. http://dx.doi.org/10.1016/B978-0-12-809372-6.00026-8

Technical Spectrum

27

Unfortunately, this concept is never clearly expressed to those entering technical careers. If understood, career choices might be better understood and surprises avoided. While all students in a particular discipline are required to take a clearly established roster of required courses, not everyone does equally well in all subjects. Correspondingly, in industry, not all technical personnel will perform well in all specialties.

Let us refer to specialties listed in Chapter 22. Note that I listed these based on most technical to least technical in terms of academic dependence. More specifically, if an engineer is working in an "analytical" group performing calculations and utilizing software, he/she is relying more on academics than those in the next category "development." Similarly, those conducting product tests are relying less on academics than those in the "analytical" group. Those in the "sales" specialty would rely least on academic subjects and so on.

This phenomena is not to imply that "sales" is easier than "analysis" or "development." But that different skill sets are required for different specialties. For example, as a "trainer," the individual must have excellent communication skills and yet understand the technology involved in the product and be able to explain the scientific principles involved. Conversely, the analytical employee should not be expected to relate well with potential customers or conduct a training course or sales presentation.

The technical spectrum ranges from more introverted behavior to more extroverted behavior thus creating somewhat of a dilemma for the "manager." Because the manager must supervise technical personnel and be responsible for their output, he/she must also communicate effectively with those who are less technically oriented, and particularly those supervising managers such as directors and vice presidents another words, up and down the ladder. This requires a personality trait I like to refer to as a "jack of all trades" and yes, maybe a master of none. I can speak to this first hand. Many technically oriented personnel find this uncomfortable largely because it is not an exact science. Managers often have to make unscientific predictions, long range forecasts, discipline personnel, and attend interdepartmental coordination meetings. Often they have to speculate on a subordinate's ability to complete a project on time and on budget. Experience is the best teacher in this regard.

The takeaway here is to match up your ability with the appropriate specialty for survival. I am certain that my brother has regrets about his experience in sales engineering. He found it necessary to undergo aptitude testing to establish a new direction for his career in which he ultimately achieved success.

Song: "Tequila Sunrise"
As popularized by: The Eagles
1985

Technical Career Survival Handbook. http://dx.doi.org/10.1016/B978-0-12-809372-6.00027-X

Job Titles

28

Conventional terminologies or titles used to describe technical personnel in order of seniority are typically as follows:

Intern
Designer
Technician
Trainer
Engineer
Principal engineer
Engineering supervisor
Engineering manager
Director of engineering
V.P. engineering

Often companies will combine these basic titles with a description to make a more specific position such as software engineer, product trainer, chemical engineer, civil engineer, or director of plant engineering. Also the terms junior or senior or I, II, III may be tacked on to describe distinct seniority levels.

Designer refers to recent college graduates with an associates of science degree (AA) and an engineer refers to recent college graduates with a BS engineering degree. Senior engineers may have several years of experience and/or an advanced degree such as a masters or PhD. Depending on the size of the company, these levels will vary considerably.

A principal engineer is often used to describe the highest technical level and often equal to or exceeding that of a manager or supervisor in salary or value to the company. Ed Gravel, a principal engineer reported to me when I was manager of development engineering at SFH. He was an important member of the staff not only for his design capability but he developed many of the patents pertaining to our products and features. He was also a "key employee," more on that will be discussed later.

Song: "Here I Go Again"
As popularized by: Whitesnake
1987

Technical Career Survival Handbook. http://dx.doi.org/10.1016/B978-0-12-809372-6.00028-1

Job Descriptions

29

Many small companies do not find it necessary to create and utilize "job descriptions" to define positions. You simply do what the boss says. However, more often, companies prepare job descriptions to clarify duties and establish salaries appropriate with positions and responsibilities.

When interviewing for a job, you may want to ask if a job description exists for the position you are seeking and if it is available for you to review. If not, perhaps it may be an indicator that the scope of duties and responsibilities are not well defined possibly putting your survival at risk.

Typically, job descriptions have some or all of the following sections (Fig. 29.1):

Title—This includes the occupational title, department name or number, date prepared, salary class, and exemption status.

General statement—This is a brief statement of the responsibilities associated with the position and to whom the employee reports.

Duties and responsibilities—These are line-by-line descriptions of what duties the position entails.

Supervision received—Describes who/how work direction is provided to the employee and how it is reviewed.

Supervision responsibilities—Who the employee is required to supervise, provide work direction and if performance reviews will be conducted.

Traits—These are the important characteristics of the ideal candidate such as honesty, self-starter, dependability, tactful, helpful, etc.

Skills—These are abilities required for the position such as oral communication, written communication, delegation, troubleshooting, planning, and organizing.

Education—Describes the minimum requirements.

Approvals—These are the personnel who approve of the job description, which may include the incumbent, his/her supervisor and a human resource department representative.

Song: "Money for Nothing"
As popularized by: Dire Straits
1985

Technical Career Survival Handbook. http://dx.doi.org/10.1016/B978-0-12-809372-6.00029-3

OFFICE JOB DESCRIPTION

Occ. Title ___MANAGER,_____ENGINEERING_____

Department _____ Division/Subsidiary _____

Date: _____ Occ. Code: _____ Salary Class: _____

Job Description Status: New ☐   Revised ☐   Supersedes: _____

Exemption Status: ☐ Non-exempt   ☐ Exempt: ☐ Executive   Administrative ☐ Professional
                                              ☐ Outside Sales

General Statement

Responsible for technical and administrative direction of design and developmental activities relating to new products. Activities include market studies, engineering analyses, prototype and product design, prototype procurement and prototype testing. Directs an engineering and product design staff sufficient to accomplish assigned functions.

REPORTS TO: Director, Engineering

Duties and Responsibilities

1. Defines, develops, assigns and periodically reviews progress of various development engineering projects in order to accomplish Division objectives. Considers validity of new development projects and defines ares of engineering improvement and cost reduction in existing products.

2. Determines the relative status of projects within department and establishes and/or revises priorities to meet desired completion schedules.

3. Communicates effectively with superior and subordinates to assure that scope, responsibility and coordination of assigned projects are fully understood. Encourages involvement of subordinates.

4. Coordinates generation of new product ideas and proposals, within the division's long range plans. Evaluates all proposals for technical merit, profit potential, patent protection and general compatibility with division objectives. Screens all product development proposals from outside sources.

5. Responsible for functioning of the department in accordance with established plans and goals. Keeps management informed as to progress and status of assigned projects, and assists in making and implementing decisions which would have an effect on overall Division operations.

6. Reviews design and proposal activities to ensure technical soundness, functional design principles, ease of manufacture, adherance to cost factors, patent protection and profit potential.

7. Provides for and establishes primary concepts and standards relating to engineering design of manufactured products.

8. Provides consultive and coordinative support to engineers on unusual technical problems, when pertinent research has exhausted available options without an answer, or when necessary to assure cooperation and assistance from other areas.

_____   _____
PREPARED/REVIEWED BY: APPROVED BY:

Figure 29.1 Typical office job description.

Salary Structure

Small companies may not have an established salary structure and simply pay what is required to hire or keep an employee. Most companies maintain a salary structure to prevent inconsistencies, provide fairness, and for budgeting. The most common salary structure is to "stair step" the salary in an orderly way usually dependent on the employees increasing value to the company. Here is a simplified example of how this works:

Salary class 10	Technician	min--------------------mid---------------------max		
		25K	30K	35K
Salary class 11	Designer	min--------------------mid--------------------max		
		30K	35K	40K
Salary class 13	Trainer	min--------------------mid---------------------max		
		35K	40K	45K
Salary class 14	Engineer, Jr.	min--------------------mid---------------------max		
		50K	65K	80K
Salary class 15	Engineer, Sr.	min--------------------mid---------------------max		
		70K	85K	100K
Salary class 16	Principal	min--------------------mid---------------------max		
		85K	100K	115K

The aforementioned annual salary rates are fictitious and are for illustration purposes only. The actual salary paid to an employee would depend on their position and their performance. In the case of a new hire junior engineer, for example, with no experience, might start at the minimum salary $50K. Thereafter, for every year of employment, the salary is adjusted based on the engineer's performance. After several years, the employee might receive a salary adjustment equating to the "mid"point of $65K. This would imply that the engineer's performance is equivalent to that of an average engineer in that company/department.

In another situation, an experienced senior engineer is beyond the midpoint at, say, $90K and receives an excellent annual review. It may be entirely appropriate to

Technical Career Survival Handbook. http://dx.doi.org/10.1016/B978-0-12-809372-6.00030-X

promote the senior engineer to principal engineer with an accompanying salary adjustment of +4% to $94K. Then as a principal engineer, the employee would be positioned
slightly below the midpoint implying that there is a considerable room for growth
under the principal engineer salary range.

Because of inflation and market competition, salary ranges are often adjusted upward
particularly during periods of high inflation, say +3%. When this occurs, employees
often receive salary increase corresponding to the movement of the range. Therefore the
revised scale for a principal engineer might, for example, look like the following:

Principal   min--------------------mid----------------------max

Engineer   87.5K   103K   118K

Often the employee's raise consists of two parts, +3% for range adjustment and
+3% for performance enhancement, totaling +6%. Or if no performance improvement is realized, the employee receives only +3%. Some companies identify the range
adjustment as a result of inflation and may present it in that way to the employee
so they recognize the difference between performance and inflation. Inflation adjustments were particularly significant during the 1980s when we witnessed double digit
cost increases.

Many companies will maintain the salary stair step data confidentially. However,
you may want to ask approximately where you are in the salary range. This will help
you determine whether you are progressing satisfactorily and if you may be in line for
a promotion to the next level. If you are topped out or near the maximum and not a
candidate for promotion, you may want to think about a job change.

To complete this topic, one should be curious as to what current salaries look like
for technical careers. A recent salary survey was compiled by the Wall Street Journal
title "What's My Major Worth." The point of the article was that if you are looking for
a high paying job, "you might want to consider some sort of engineering."

Based on 2016 salary data, Table 30.1 shows the results of a Payscale, Inc. survey
but bear in mind the salaries listed are based on *mid-career median levels*, which
equate to about 20 years of experience. With the exception of petroleum engineering,
the results look realistic for median pay.

Table 30.1 Early versus Mid-Career Pay

Rank	Major	Degree Type	Early Career Pay	Mid-Career Pay
1	Petroleum engineering	Bachelor's	$101,000	$168,000
2	Nuclear engineering	Bachelor's	$68,200	$121,000
4	Chemical engineering	Bachelor's	$69,500	$118,000
5	Electronics and communications engineering	Bachelor's	$65,000	$116,000

Table 30.1 **Early versus Mid-Career Pay—Cont'd**

Rank	Major	Degree Type	Early Career Pay	Mid-Career Pay
6	Computer Science (CS) & engineering	Bachelor's	$69,100	$115,000
7 (tie)	Electrical & Computer engineering (ECE)	Bachelor's	$67,000	$114,000
7 (tie)	Systems engineering	Bachelor's	$67,100	$114,000
9	Aeronautical engineering	Bachelor's	$65,100	$113,000
10 (tie)	Computer engineering	Bachelor's	$68,400	$109,000
10 (tie)	Mining engineering	Bachelor's	$71,500	$109,000
12 (tie)	Electrical engineering (EE)	Bachelor's	$66,500	$108,000
12 (tie)	Mechanical and aeronautical engineering	Bachelor's	$61,100	$108,000
14 (tie)	Aerospace engineering	Bachelor's	$64,800	$107,000
14 (tie)	Computer science (CS) and mathematics	Bachelor's	$62,900	$107,000
18 (tie)	Materials science and engineering	Bachelor's	$64,600	$105,000
20 (tie)	Mechanical engineering (ME)	Bachelor's	$62,500	$102,000
26	Industrial engineering (IE)	Bachelor's	$62,800	$99,600
29	Structural engineering (SE)	Bachelor's	$57,500	$97,700
32	Software engineering	Bachelor's	$62,500	$96,800
33	Electronics engineering	Bachelor's	$60,700	$96,700
34 (tie)	Biomedical engineering (BME)	Bachelor's	$60,900	$96,400
34 (tie)	Engineering management	Bachelor's	$63,100	$96,400
38	Civil engineering (CE)	Bachelor's	$55,600	$94,500
39	Manufacturing engineering	Bachelor's	$60,400	$94,000
41	Environmental engineering	Bachelor's	$52,400	$93,400

PayScale, Inc.

Song: "Money"
As popularized by: Barrett Strong
1960

Key Employee

31

In a competitive market place, it may be extremely difficult to fill open technical positions without radically deviating from the salary structure, which may already be in place. It is not practical to revise the structure based on hiring an individual employee for a unique position. Therefore, some companies incorporate a category for these unique situations often referred to as "key employee," which are essentially a unique set of benefits. Earlier I mention a position for a senior chemist was created to analyze and recommend waste stream treatment processes and equipment. In this example, the candidate could have been offered a key employee status as an added inducement to accept the position. I was not privy to the details of his or her offer of employment. The key employee status is basically a means of offering incentives or rewards to employees without disrupting the salary structure.

What are the advantages of the key employee status to the candidate? There are several and may vary among corporations. The most common features are of the following:

- additional vacation days
- shares of company stock or phantom stock
- relocation expenses
- additional life insurance
- company car
- country club membership

I was offered a key employee position after interviewing for an engineering manager position for which I accepted. The additional vacation days appealed to me as I was only eligible for 2 weeks with my previous company. Additionally, I received phantom stock later after being promoted to director of engineering. A predetermined block of phantom stock (aka time-vested stock shares) is issued to an employee for which he or she receives the dividends but does not actually take possession of the shares until a specific period of continuous employment with the company is realized, perhaps 5 or 10 years. This enhances your survivability and will instill loyalty to the company in a similar way that a pension does.

Song: "Sharp Dressed Man"
As popularized by: ZZ Top
1992

Technical Career Survival Handbook. http://dx.doi.org/10.1016/B978-0-12-809372-6.00031-1

Nondisclosure Agreement

32

When you become an employee, part time, or full time, you will likely be privy to a vast amount of confidential and proprietary information associated with the company's operation. Some examples include the following:

Sales and marketing strategies
Client or customer lists
Rates and pricing structure
Patents pending
Engineering drawings
Computer software
Test results
Tools
Specifications
Trade secrets/processes
New products or features

Because the company values the previously mentioned information, they will require that their employees deem these items confidential. Therefore, odds are that you will be requested to sign a nondisclosure agreement (NDA) at some point along your technical career path. It may also be referred to as a confidentiality agreement, confidential disclosure agreement, proprietary information agreement, or secrecy agreement.

These agreements are *legal contracts* between the employee and the company that outlines confidential material, knowledge, or data that the parties wish to share with one another for certain purposes but wish to restrict access to or by third parties. It is a contract through which the parties agree not to disclose information covered by the agreement. An NDA creates a confidential relationship between the parties to protect any type of confidential and proprietary information or *trade secrets*. As such, an NDA protects nonpublic business information.

Here are some of the main requirements of a typical NDA:

1. A specific time period is established for the agreement usually at or near the employee start date through the time of employee termination plus 1 year.
2. Specific types of company confidential information may be listed.
3. Company confidential information which the employee is privy shall be surrendered upon termination.
4. The employee shall be prohibited from recruiting, contacting, or soliciting the company's employees for a designated period upon termination.
5. Data or information the employee brings to the company when hired may be specifically identified.

Technical Career Survival Handbook. http://dx.doi.org/10.1016/B978-0-12-809372-6.00032-3

NDAs and associated legal requirements may vary not only among corporations but from state to state. You may want to contact an attorney knowledgeable in the employment field if you foresee conflicts. Court battles in this area are rare and I have not heard of specific cases. But I have seen a red flag raised when an employee terminated employment and was immediately hired by a direct competitor in a similar capacity.

Song: "Open Secrets"
As popularized by: Rush
1987

Company Size

33

While the legal definition of company sizes varies by country and industry, for sake of discussion, it is generally agreed that in the United States:

Small companies are less than 125 employees.
Medium companies are 125–400 employees.
Large companies are 400–1000 employees.
Mega companies are 1000 plus.

In my experience, small companies in the technical arena are most common in engineering consulting, research, and development rather than the manufacturing field. Although I was employed by an air compressor manufacturer with fewer than 100 employees but their product line was primarily manufactured by the parent company in Germany. The three main difficulties that *small* technological companies face are of the following:

1. Having sufficient capital to cover operating expenses and payroll particularly during downturns in the economy. Insurance costs, liability, health, and life are extremely high in the United States.
2. Difficulty in getting products or services into the marketplace either by direct sales, distributors or through manufacturer's representatives. Some utilize all three methods plus offering their products on an original equipment manufacturers (OEM) basis.
3. The ability of an expert in the technical field to run a small business. There is often a failure to distinguish between small business manager as an entrepreneur or capitalist. While nearly all owner–managers of small firms must assume the role of capitalist, only a minority of these managers will also act as entrepreneur.
4. Transitioning from a small start-up organization that does not dominate the market to medium size company competing in a national marketplace. This may necessitate a significant company restructuring.

Medium size companies usually separate management from the owners and emerge in the marketplace as a significant threat to the large companies that dominate. Many are privately held by a small number of stockholders. These medium size firms gain access to capital markets and generally expand into national or international operations. While I was employed by a medium size company, SFH, we penetrated the marketplace aggressively through direct sales, distributors, manufacturer's representatives, licensees, and OEM accounts both nationally and internationally. Like large companies, medium size companies tend to be efficiently structured and have well-established policies, procedures, and salary structure.

Large and mega companies are legal entities or corporations with many owners who are called stockholders. These corporations can issue stock to raise capital, elect officers, file articles of incorporation, pay stockholders dividends, and acquire other

Technical Career Survival Handbook. http://dx.doi.org/10.1016/B978-0-12-809372-6.00033-5

companies. I worked for three mega corporations and because they represent a major portion of the technological workforce, they also experience the greatest need for engineers and technicians.

The advantages I saw working for large/mega corporations were of the following:

• opportunities across various departments for advancement
• regularly scheduled training and development programs
• competitive salary and benefits
• reimbursement for seminars and college courses
• well-established products and technology.

One could argue that employment for engineers and technicians in large/mega companies and for that matter, medium size companies is not as certain as in past years. But regardless, I believe there will always be unavoidable changes in company stability as influenced by economic, political, and energy utilization activity. However, large/mega companies are more likely to exercise acquisitions, spin-offs, and mergers than medium and small companies thereby affecting survivability. An affected employee may consider this a plus if it results in a promotion or a better opportunity in the future. More on this is discussed in Chapter 51.

Song: "Shooting Stars"
As popularized by: Bad Company
1977

Company Organization

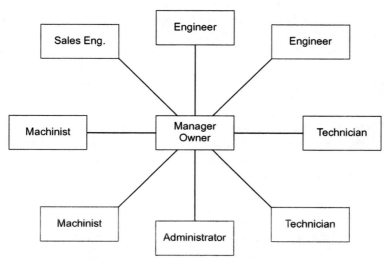

Regardless of a company's size in terms of head count, there are several ways in which they can structure their workforce. Although in small start-up companies, it is not unusual to find everyone reports to the boss/manager/owner as in a *wheel organization* (Fig. 34.1).

This is a convenient, rapid response approach when there are no actual departments comprised of multiple employees. There is no attempt to create a measurement or level of responsibility between employees. Also there is reluctance on the part of employees to act without consent of the boss. When the head count exceeds a dozen or so employees, it is time to consider alternative structures.

The most common organization is the line organization (Fig. 34.2).

This organization structure is utilized in small, medium, and large technological companies. The line organization diagram shows levels of authority or seniority vertically (chain of command) and simultaneously shows peer or similar ranking horizontally. Line organizations tend not to be broad horizontally as there is usually a practical limit to the functions or departments required, i.e., engineering, manufacturing, sales, administration, etc. Whereas vertically, the experience of subordinates will determine the number that can effectively report to the supervisor. In other words, less supervision is required for senior level personnel.

Figure 34.1 Wheel organization.

Technical Career Survival Handbook. http://dx.doi.org/10.1016/B978-0-12-809372-6.00034-7

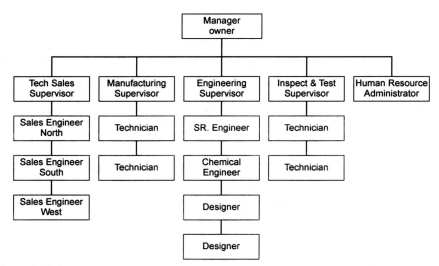

Figure 34.2 Line organization.

Medium and large companies will have line organization charts for each department, i.e., engineering, marketing, sales and manufacturing, and so on. Collectively, they define the company organization structure. Actual names would normally be inserted in the diagrams (Fig. 34.3).

This is an alternative organization structure that focuses more on the tasks or projects usually associated with products that a company produces. The primary advantage of this *product/project organization* is that members of each product line are simply focused on the success of the single product and attention is not diverted to other lines. Thus, team members have specialties and backgrounds associated with the specific products and are hence experts. There is also a competitive atmosphere, which is created between the product lines that inherently leads to successful results. On the negative side, unless there is sufficient cross training, the departure of a team member may create a serious void and disrupt progress of that product marketing effort.

Product/project organizations tend to be broad horizontally because both products and projects may be numerous. ITAC, a consulting engineering company that I worked for part time for at one point had as many as 11 departments reporting to the president. One department consisted of 21 employees under a single manager. Again, because consulting engineering companies tend toward employing more experienced personnel, less supervision per employee is required.

A *hybrid organization*, sometimes referred to as a matrix organization, is created when the line organization is combined with the product/project structure on a temporary basis. This is typically used for consulting engineering companies where multiple projects are encountered in the normal course of business. This structure requires a project manager be assigned for the duration of the project, let us say 6 months. For the project duration, a mechanical engineer, mechanical designer, a civil engineer, and a civil designer are temporarily assigned to the project. While working on the project, the team members continue to report to their respective discipline supervisors/

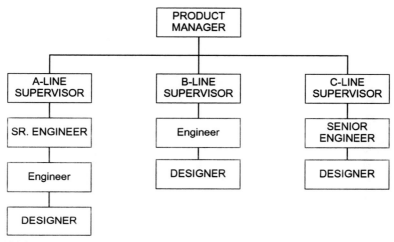

Figure 34.3 Product organization.

managers although they receive project-related work direction (dotted line) from the project manager. This requires the project manager, sometimes referred to as the program manager, has an understanding of the team member's capabilities and a reasonable expectation of their abilities to complete the work. At the conclusion of the project, the discipline managers should collaborate with the project manager to assess the job performance of the team members in preparation for their annual performance reviews.

The key to success with the hybrid organization is team members must be self-starting and require a minimum amount of supervision in their discipline. Normally, there is no need to move personnel work spaces within the office for the sake of the temporary assignment.

Song: "Never Know"
As popularized by: Jack Johnson
2008

Pensions

Chapter Outline

Company pensions were paid as far back in history as 1935 when the federal government established the railroad retirement system to force companies to meet their obligations to employees. Since then, companies introduced pension plans to be competitive in attracting workers. Ford motor company did not start offering hourly employees pension plan until 1950. However, most companies now offer 401K plans as an alternative to pension plans.

What Is a Pension?

A traditional pension plan is designed to provide an income source or annuity for employees in retirement after years of service with the company. The monthly rate may vary with the type of annuity but is directly proportional to the years of service and salary at the time of retirement. Continuous service may be as few as 3 but often as many as 10 years to qualify depending on the company plan.

What Are the Types of Pensions?

An *ordinary life annuity* provides a fixed monthly benefit that is paid to the retiree for the rest of their life.

A *guaranteed dollar value annuity* will be paid to the retiree for the rest of their life. If the guaranteed dollar value is not exceeded during that period, the remainder of the amount will be paid to the designated beneficiary until the value is paid out. At that time the payments will cease.

A *certain life annuity* provides a fixed monthly benefit that is paid to the retiree for the rest of their life. If the retiree dies before a specified number of months, the remainder of the amount will be paid to the designated beneficiary until the specified number of payments have been made. At that time the payments will cease.

Technical Career Survival Handbook. http://dx.doi.org/10.1016/B978-0-12-809372-6.00035-9

A *joint and survivor life annuity* provides a fixed monthly benefit that is paid to the retiree for the rest of their life. Upon their death, a designated beneficiary will receive the payments until the time of their death. At that time the payments will cease.

A *lump sum* pension is not an annuity and is a calculated present value of the retirement benefit based on their actuarial life expectancy.

What Companies Offer Pensions?

Publically owned companies such as those on the Fortune 100 list are more likely to offer pension plans than those companies in the private sector. According to the Bureau of Labor Statistics in 2012, only 9% of private employers offer pension plans. Whereas 49% of large employers, over 500 employees are likely to offer pensions versus only 8% for small companies. Utility, financial, and information service companies are likely to offer pension plans for their qualified employees.

What Is an Example of an Annuity Payment?

One Fortune 100 company uses the following formula:

1.6% × pension years × final monthly pay (less SS) = monthly pension payment.
0.016 × 30 years × $4000/month = $1920/month.

I would not recommend bringing up this topic during a job interview as it might jeopardize your chances of receiving an offer. Save it for later at the point you are discussing salary and benefits associated with an offer.

Song: "A Hard Day's Night"
As popularized by: The Beatles
1964

Training

36

In previous Chapter 19, I emphasized the importance task of the manager to provide training to subordinates as a pathway to their future promotion. However, I discovered that this principle may not always be looked upon favorably. I hired a "tree expert" to remove some dangerously large tree limbs overhanging a very old house I purchased several years ago. I noticed when the crew arrived at my house, the boss/owner climbed the tree with spikes, ropes, and chainsaw while the younger subordinates were busy with the low-risk groundwork gathering cut limbs, raking, and feeding the chipper. So after he descended from the tree, I asked him why he had not trained the young bucks how to climb and to do the risky work? His response was "oh no, then they would quit and start their own business in competition with me." I thought that perhaps this was the exception to the rule.

Engineering organizations, like other businesses, are constantly keeping up with technology. This is achieved through webcasts, mentoring, memberships in societies, tuition reimbursement, and paid training. There are four categories of technical training that I have personally experienced.

My first job immediately upon graduation was with Hamilton Standard Division (HSD) of United Technology as an entry level design engineer. I was assigned to the environmental controls system department during the aerospace boom. The products were aircraft and space vehicle components. HSD required new hires to participate in a 6-week partial day *company training program* referred to as the manufacturing training program consisting of both classroom and hands on shop machinery work. At the conclusion of the course, I had manufactured, assembled, and tested a small pressure relief valve utilizing many different types of machinery found in a basic shop. Machining required for the project consisted of lathe turning, drilling, boring, milling, and grinding and fortunately no first aid was required.

The most common, particularly in small companies is *on-the-job* training (OTJ).

My second job was with General Electric Gas Turbine Division where I was hired as a replacement for a senior engineer who was transferring to another department. I was fortunate as he was a great mentor for the short time I had to come up to speed in the position of control components engineer. I was responsible for design, vendor selection, and test of gas turbine control components such as hydraulic pumps, actuators, valves, and atomizing air compressors. My OTJ training lasted for approximately 6 weeks before I was completely on my own.

Another type of training would be considered *industry-required training* such as what Dominion Virginia Power required that I would be qualified for nuclear design projects. The training involved 6 months part-time Nuclear Regulatory Commission (NRC) "basic engineering training" that was quite rigorous and necessitated passing grades on mandatory tests. Subjects included chemistry, physics, heat transfer, and thermodynamics.

Technical Career Survival Handbook. http://dx.doi.org/10.1016/B978-0-12-809372-6.00036-0

The fourth type of training and certainly the most common encompasses various subjects as determined by the management of the company which I refer to as *elective specialties training*.

This training may be conducted for a group of employees as led by a designated trainer as in a classroom fashion, individually online or at a remote location. I have experienced all three formats but realize that the company/department budget is a major factor in determining what training programs are viable.

Regarding specialties training, while with an engineering consulting company I volunteered to enroll in a 3-day seminar offered in Texas to study CAESAR II. This is software that is used to evaluate forces, moments, and stress levels in the design of liquid and gas process piping. While I did not realize it at the time, this training provided me with a skill set that proved invaluable while working for consulting engineers and later on my own.

In addition to CAESAR II, there are many types of software used by engineers to facilitate design work. This software is typically licensed by the user according to the number of "seats." Here are a few others shown in Table (36.1):

Most states require continuing education units to maintain a valid professional engineering license necessitating enrolling in approved courses. I will address this in a later section.

While everyone generally agrees that training is a positive, there are common obstacles preventing it in a technical environment.

- Training should be relevant to the job. Will it enhance the employee's performance now or in the future? Will it be applicable frequently? This is usually a judgment call by the manager approving the training.
- Is the cost of training in the budget? It is good practice to consider the cost of seminars, professional training, and Internet training in the annual department budget. Be logical in determining who will participate.
- Avoid the boredom factor when selecting training programs. Perhaps spread the training over lunch time periods. Be certain the trainer is well familiarized with the subject and can hold the participants interest.
- Be certain work schedules allow sufficient time to devote to training. How will the lost time of personnel to training be covered by others? Obviously, select time periods when knowingly the work load will be reduced.

I highly recommend bringing up the topic of training during an interview. It is important to understand what avenues for growth through training exist within the

Table 36.1 Common Engineering Software

Software	Calculation	Discipline
ANSYS	Heat transfer analysis	Mechanical engineering
STRUDL	Structural analysis	Civil/structural engineering
Pspice	Circuit analysis	Electrical engineering
CHEMCAD	Process analysis	Chemical engineering

company. You will also convey the message that you are personally interested in utilizing training to survive and possibly be promoted within the company.

Song: "Teach Your Children"
As popularized by: Crosby, Stills and Nash
1970

Benefits

Chapter Outline

Company benefits, sometimes referred to as "perks," are offered at little or no cost to employees as compensation in addition to salary. In some instances, companies may have waiting periods of 60 or 90 days before employees are eligible for particular perks. Benefits are offered by companies in order to:

1. Maintain competitive compensation packages in their respective industry
2. Attract high quality candidates within their industry
3. Create a favorable morale among the employees
4. Establish incentives for employees as they advance within the company

The growth in benefits offered by employers in technical industries in the last 20 years has been significant. Prior to that time, benefits only included life insurance, vacation and holidays, health insurance, and retirement plans. Items 1 and 2 mentioned were the primary reason benefit programs were expanded.

Benefits consist of two types, mandatory and optional. The former are those that are *required* by law include the following:

Social Security

It requires a payroll deducted amount from the employee with a future amount to be paid by the federal government at a retirement year or at a time when the employee becomes disabled.

Worker's Compensation

It provides wage replacement and medical benefits to employees injured on the job.

Technical Career Survival Handbook. http://dx.doi.org/10.1016/B978-0-12-809372-6.00037-2

Unemployment Insurance

It requires a payroll-deducted amount from the employee with a future amount to be paid by the federal government should the employee lose their job for a specific period of months.

Obamacare

The firms with 50 or more employees not offering health-care plans will be required to contribute to the employee's health care costs if subsidized by the government.

Optional benefits come in all sizes and shapes. They all come at a cost to the employer and are a part of the employee's total compensation. The Bureau of Labor Statistics published a list of the most common optional benefits in July 2013. The top 10 from most to least common are as follows:

Unpaid family leave
Vacation pay
Holiday pay
Health/medical cost
Prescription drugs
Paid jury duty
Paid sick leave
Paid funeral leave
Paid retirement plan
Life insurance

One of the most recent optional benefits to be offered is the ability to telecommute. This is particularly desirable with women who have small children or those who must drive considerable distance from home. More on this will be discussed later.

Unusual optional benefits currently offered by a few companies include the following:

Travel expense coverage during vacation
Payment for "Life Coaches"
Wellness programs including messages, acupuncture therapy, and personal trainers
Payment for housecleaning and errands
Tuition reimbursement.

Song: "It's Five O'Clock Somewhere"
As popularized by: Allan Jackson and Jimmy Buffett
2003

Public Versus Private Technology Companies

Chapter Outline

Undoubtedly in the course of a technical career, one will consider employment with either a private or public company not to be confused with a governmental employer. At first glance the differences may not be obvious. Here is a breakdown of the main differences:

Private Companies

- They are privately held or owned by private investors, the company founder(s), family, or a private management company.
- The company's finances may be kept secret and not shared with the employees.
- The company's financial success or failures may or may not directly affect employees.
- Must seek private funding to expand their business or purchase inventory.
- Management has independence from shareholders.
- Often morph into public companies with continued growth and profitability.
- Management often employ family members or friends and treat them more favorably over other employees.

Public Companies

- They are owned by stockholders who purchased an initial public offering or purchased existing stock from the stock market that was previously owned by stockholders.
- Owners or stockholders have a claim to the company assets and profits but their stock value fluctuates depending on the company's financial performance and perceived value.
- The company's performance is in the public domain and available to employees and stockholders.
- May raise capital in financial markets by selling stocks or bonds to expand their business or purchase inventory.
- Management must answer to shareholders and file disclosure statements to the Securities Exchange Commission.
- Seek to maximize stockholders profits and reflect that in the employee's compensation.
- Often buy or sell new related businesses using corporate profits.

Technical Career Survival Handbook. http://dx.doi.org/10.1016/B978-0-12-809372-6.00038-4

Having worked for two private companies, I found that the secrecy of the business was unsettling particularly not having a clear vision of the company's future. Working for public companies eliminates many of the mysteries behind the company's goals, objectives, and financial position thus improving your chances for survival.

Song: "Bad Company"
As popularized by: Bad Company
1974

Industries

The industrial revolution of the 18th and 19th centuries resulted in a transition from hand tools to machines; from wood fuel to biofuels and coal; and from man and animal power to water and steam power. The second industrial revolution in the 19th century was marked by the introduction of steel production, electrification, mass production, railroads, and petroleum and chemical production.

That era gave way to industries such as:

Iron and steel
Railroad
Machine tools
Petroleum
Chemical
Maritime
Rubber
Plastic
Automobiles
Engines
Turbines
Telecommunications

These industries grew as a result of engineers, scientists, and technicians and their never-ending efforts to optimize manufactured goods. The growth of these industries was a combined effort of many disciplines working together toward a common goal. Often people associate a particular industry with a single discipline. For example, they believe that chemists and chemical engineers are solely responsible for advances in the chemical industry, or electrical engineers are solely responsible for advances in telecommunications industry. This is not the case at all. I was amazed at how many mechanical and civil engineers and chemists were employed at both fossil and nuclear power plants. Similarly, in the chemical industry there are many electrical, civil, and mechanical engineers providing on-site plant design and maintenance support.

The lesson here is, do not attempt to judge the type of jobs required by a specific industry. As you enter the job market, you will become aware of the great number of opportunities available in industry for technically skilled personnel. Recently a friend told me that his son, a mechanical engineering major, was applying for an open position at Capital One, a major financial company.

Song: "Workin' In a Coal Mine"
As popularized by: Lee Dorsey
1966

Technical Career Survival Handbook. http://dx.doi.org/10.1016/B978-0-12-809372-6.00039-6

Industry Trends

Often overlooked in the media are trends that have developed in industry over many years. Having been in the workforce for decades and working in several different industries, there are obvious trends that have occurred to me, and I believe it is important to recognize them and how they might influence one's technical career. Here are a few of my past observations in that regard:

Aerospace: Starting around the mid-20th century, commercial aircraft manufacturing accelerated with development of high capacity, long haul planes such as the Boeing 727, 737, and 747. Advances in materials, machining, and engineering analysis made this possible. Engineers were hired in droves not only by airframe manufacturers such as Boeing, Lockheed, and Grumman but also by hundreds of component providers all over the country. Military aircraft, helicopter, and missile development and production ramped up particularly during the Vietnam War era.

Space exploration: From the mid- to the end of the 20th century, space exploration programs flourished largely due to the governmental promotion and funding. While I was employed as a design engineer at Hamilton Standard Division, working on the Boeing 747 environmental controls, there was a large group of engineers on the floor below working simultaneously on moon exploration backpacks. Space exploration began to wind down at the end of the century primarily due to the de-emphasis by the US government.

Superfund: In 1980, legislation was passed with mandate for cleanup of many sites in the United States that were considered contaminated with hazardous chemicals referred to as "pollutants or chemicals." Under the authority of the Superfund, the Environmental Protection Agency was formed to manage the design, construction, and operation of clean-up facilities, thus creating a huge industry comprised of engineers, scientists, and technicians. Collectively they would analyze the waste, prepare specifications, select equipment, and oversee the installation and operation of the remediation processes.

Alternative fuels: Starting in the 1970s with the infamous Arab embargo, fuel supplies have made the headlines to this day. From my days with General Electric, Gas Turbine Div., I recall that Distillate was the standard fuel in the industry. Then came the Arab embargo and from that point on there was no standard. At first, natural gas looked promising but subsequently, the scarcity made it less desirable. Then heavy fuel oils looked promising but required elaborate cleaning systems to make them suitable for burning. Alternative fuels such as blended ethanol/petroleum also hit the market showing promise but resulted in higher production costs. Finally, based on fracking/water injection, natural gas and oil production once again appear promising economically, but at the same time is considered unfavorable by environmentalists.

Alternative energy sources: As far back as the 1970s, companies have invested in geothermal wells. While at Sundstrand Fluid Handling, my engineering group was involved with the development of "downhole" pumps that inject water as far down as miles, then extract the steam that is created by the hot earth to expand through a turbine and produce electrical energy. Solar energy continues to be exploited not simply by placing collectors on building roofs but also by launching collectors into space and beaming power back to earth. The latter concept has been confirmed, but the economics will require optimization.

Technical Career Survival Handbook. http://dx.doi.org/10.1016/B978-0-12-809372-6.00040-2

High tech: In the mid-1940s a team of Bell Labs scientists invented the point-contact transistor that could amplify electrical signals with the intent of using them for telephone signals. The device replaced vacuum tubes that were unreliable, required too much power, and emitted excess heat. This invention led to the development of a class of semiconductor amplifiers used in the manufacture of electronic devices since the mid-1950s beginning with the "transistor radio." Since then the high-tech industry flourished with the production of modern devices, such as cell phones, television, medical instruments, recording devices, robotics, and computers.

Green Jobs: Combining the concern for fuel cost and availability and protecting the environment, we are currently seeing an emerging market focused on the production of high-efficiency machinery, low-emission power generation, and alternative energy sources. The practicality of newly introduced solutions, that is, the cost versus benefit, remains to be determined presenting tremendous challenges to scientists and engineers. The economics once again becomes clouded by the influence of government subsidies. More on this subject in the next section.

So how do these trends impact technical careers? Obviously, they represent career opportunities for engineers, scientists, and technicians of all disciplines. Having personally experienced three of the above trends first hand, the lesson learned is to understand: What is the impetus for an industrial trend? Are legislation and government funding in place? How sustainable is the trend? Should you jump in? Will you survive in the long run? Be aware that government funding can often change without much advanced notice. Of the seven trends listed, high tech has been least affected by politics and government, thus far. Be on the lookout for the next industry trend and maybe you will ride the wave.

Song: "Man On the Moon"
As popularized by: R.E.M.
1992

Green Jobs

Disclaimer: This subject matter is not necessarily treated with political correctness. As radio commentator Rush Limbaugh points out, all things are political, but I might add, some more than others.

In 1970, Congress established the Clean Air Act after a fire broke out on the Cuyahoga River in Cleveland, Ohio during 1969. The first Earth Day was organized in 1970 emphasizing environmental awareness. In the previous section, I mentioned that during 1980, legislation was passed with mandate for cleanup of many sites in the United States that were considered contaminated with hazardous and radioactive chemicals, that is, a "brown field," such as Hanford WA that I was involved with. The intent of those cleanup projects was to restore the sites to a noncontaminated condition. Hence, the term "green field" was coined although we did not refer to our remediation work as a "green job" at that time. The term was coined later.

What followed over the next several decades was the emergence of an environmental movement that produced a wave of books, media coverage, conferences, symposia, and political dialog. The movement then steered toward global warming, resource depletion, and extinction of the species. Also green became the buzz word associated with materials, processes, automobiles, buildings, recycling, energy, and many other goods and services. Then in the fiscal year 2010, The Bureau of Labor Statistics (BLS) received funding to define and produce data on green jobs. The term "green jobs" has crept into the lexicon and has now become very familiar to those who keep up with the environmental news. Green jobs are defined by BLS as being either of the following:

- "Jobs that produce goods or services that benefit the environment or conserve natural resources."
- "Jobs in which workers' duties involve making their establishment's production processes more environmentally friendly or use fewer natural resources."

Before discussing the BLS initiative, it is important to understand the role of engineers, scientists, and technicians in relation to the development of processes, products, and services that people want, need, and desire. In all cases, they begin with design criteria. Let us consider a power plant that generates electricity. How will the plant be designed? Basic questions have to be resolved such as:

Where to locate the plant?
Which fuel is to be burned?
How many megawatts of power are to be produced?
Which power plant technology is to be used?
What should the plant cost?
What would be the method of cooling?

Technical Career Survival Handbook. http://dx.doi.org/10.1016/B978-0-12-809372-6.00041-4

The answer to all these questions must be based on the most reliable, economical, and available technology that exists *at the time*. Then, answers to these questions become constraints and part of the design basis for the plant. Design work begins, earthwork ensues, materials and machinery are ordered, delivered, and installed, and commissioning begins. Therefore, many years after the plant has been operational, it is unreasonable to criticize the power plant design because it produces much pollution, is inefficient, or utilizes excessive cooling water. That is pure hind sight. It was designed and approved based on the criteria and the best technology available at the time.

This principle is true for all engineered products. Take modern electric automobiles, for example. The design basis decision concluded that an electric motor would be the propulsion means. On the surface, this circumvents the fact that the electricity must be (1) produced in a power plant that likely burns fossil fuels and (2) delivered to a charging station. Therefore the cost of operating the vehicle looks very attractive from a cost-per-mile perspective, and the car does not emit pollutants. Further, the cost of manufacturing the car becomes secondary in the design basis if tax credits are available thereby seemingly lowering the vehicle cost making it competitive with fossil fuel–powered vehicles.

OK, so how do design criteria and green jobs relate to the BLS initiative?

First, the purpose of the initiative is to determine:

- the number of and trend over time in green jobs;
- the industrial, occupational, and geographical distribution of these jobs;
- wages associated with these jobs.

Second, where are the so-called green jobs found? They are found in:

- production of green goods and services;
- the processes and practices that are environmentally friendly.

Therefore, if you are focused on landing a green job, for whatever reason, you might want to consider companies producing products associated with solar, wind, wave, geothermal, biodiesel, fuel cells, mass transit, hybrid vehicles, green buildings, air quality, waste management, pollution reduction, recycling, and many other such areas. In 2011, there were 3.4 million green goods and services (GGS) jobs, accounting for 2.6% of total US employment. While this sounds impressive, effectively BLS has created a category that is just another way of accounting for employed engineers, scientists, and technicians, much like tracking diversity in the workplace or in the classification for professional engineering positions. Do not look for a surge in employment advertisements for green jobs. Like any job opening in our capitalistic society, they are usually a result of a market opportunity due to a better price, delivery, product, or performance. Also, recall what I highlighted in the previous section: If government funding or subsidies are associated with the product or process, be cautious for your career survival.

Song: "Big Yellow Taxi"
As popularized by: Joni Mitchell
1970

Regional Manufacturing Centers

42

According to a Brookings Institute study, May 2012, manufacturing constitutes about 61% of the value of US exports, 11% of our GDP and 66% of our R&D spending. Therefore it is important for engineers, scientists, and technicians to understand where these manufacturers are located. The study determined that the majority of the manufacturers are located in metropolitan areas, thus providing the necessary workforce. Here is a list of the "anchor industries" groupings that the study evaluated. Also listed is the number of metropolitan areas where they are located (Table 42.1).

Another consideration important to engineers, scientists, and technicians is compensation. While manufacturing wages tend to be higher than other industries, wages vary widely from area to area. In 2010 most metropolitan area manufacturing plants paid slightly less than the $55K, on average.

The study determined that during 2000 to 2010, all regions lost manufacturing, while during the last two years the Midwest posted the strongest gain in jobs. Surprisingly, most plants tend to be small, that is, less than 60 employees. These smaller plants are responsible for designing, developing, and producing an increasing amount of the content of manufactured goods.

You may find that these industries are regionalized which may allow you a certain amount of job movement within a particular region, thereby improving your survivability. By that, I refer to the fact that you will likely find other related employers nearby who will value your experience should you decide to change jobs.

Table 42.1 Manufacturing Metropolitan Areas

Industry	Number of Metropolitan Areas	Primary Location
High-tech, computers and electronics	35	North and West United States
Low-wage manufacturing	39	East United States
Chemicals	50	North and East United States
Machinery	60	East United States
Food	50	East United States
Transportation equipment	90	East United States

Brookings Institute 2012 data.

Song: "Kansas"
As popularized by: Wilbert Harrison
1959

Technical Career Survival Handbook. http://dx.doi.org/10.1016/B978-0-12-809372-6.00042-6

Foreign-Owned Companies

Chapter Outline

Personal Concerns 110

Why are foreign-owned companies important to those seeking a technical career? Several of the reasons are highlighted in a June 2014 report of a joint project of the Brookings Institute and JPMorgan Chase. They determined that about 5.6 million workers in the United States were employed through foreign direct investment (FDI) across every sector of the economy. This number rose from 1990, peaking around 2000, and then falling off. Nearly three-quarters of these jobs are located in the 100 largest metropolitan areas. The average large metro area contains FDI-type jobs from 33 different countries and 77 different city-regions worldwide.

FDI jobs are largely concentrated in the nation's technology- and skill-intensive advanced industries; however, they have also become more services oriented in recent years. This amounts to 5.5% of private employment in the average large metro area ranging from over 13% in Bridgeport CT to as low as 1% in Provo, UT. Most FDI companies were established through mergers and acquisitions and not as a direct source of job creation.

One of the two FDI companies I worked for manufactured *industrial air compressors* sold in a variety of markets such as auto repair shops, manufacturing plants, food processing, printing, and chemical plants, and many others. The company started by shipping air compressors from their plant in Germany to the United States where they warehoused, tested, and shipped them to distributors across the United States. Over the years, they found a niche for a reliable heavy duty air compressor that was easy to service and maintain but not necessarily low priced. Over time, they became successful in capturing a respectable share of the market and later branched out into other related product lines such as air filters, air dryers, and moisture separators. In an effort to reduce their labor cost, they chose to locate in populated but a non-metro area of Virginia.

Another FDI company, Adtechs, I was employed by as a consulting mechanical engineer, was based in a large metro area of Virginia for two reasons: they wanted the convenience of a major airport, and close proximity to lawmakers and Environmental Protection Agency personnel. Their presence in the United States was largely due to the Superfund era/trend that I described in an earlier section. The company had substantial experience designing and constructing *hazardous and radioactive waste treatment* facilities in Japan and so they saw a niche in the United States. Technologically, they

Technical Career Survival Handbook. http://dx.doi.org/10.1016/B978-0-12-809372-6.00043-8

were successful but were not well versed in project management and cost control. A major processing facility they constructed at Hanford WA was bid at $28 M but ended up coming in at $64 M. After only 7 years in the United States they dissolved their operations.

Personal Concerns

1. From a personal perspective, my takeaway working for FDI companies is the "us versus them" syndrome that pervades the work force. There is always the concern that the foreign management is not in step with the wants, needs, and desires of the US management and workers, largely due to the communications. It would seem that it is not sufficient to have monthly joint progress and strategy meetings; they lack the daily face-to face communications. This creates a morale problem that is difficult to overcome, but has improved largely through the many forms of modern communication.
2. There are also the cultural challenges. Of particular difficulty for me was working for a Japanese manager who, although technically quite competent, had difficulty communicating his directions and scope of work. Of course, things did improve significantly with time.
3. I believe my input was less important during my years employed by FDI companies. Perhaps during current times, engineering management has placed more emphasis on employee feedback and in fact it may vary considerably among FDI companies. The automotive industry has emphasized "quality control circles" among factory workers and certainly this has spilled over to technical professionals as well.

I would not discourage anyone from working for an FDI company. If you are considering employment with them, possibly talk to current employees in order to determine if you are a good fit and can survive in that organization.

Song: "Back in the USA"
As popularized by: Chuck Berry
1962

Manufacturing

Since a great many technical personnel will likely spend a portion of or perhaps their entire career in manufacturing, a large sector of the economy, let us look at the characteristics of this environment.

The main parameter for a manufacturing company is *shipments*. They are generally tracked on a weekly and monthly basis and measured against the shipment forecast. The *sales forecast* is prepared at the end of the prior year and often extrapolated from the results of the previous year. Taken into account are the level of buying activity expected in various markets, penetration into new markets, and new products that are expected to be sold into either existing or new markets. The forecast numbers are submitted by the sales engineering department and crunched and manipulated by the marketing department. The sales forecast is then reviewed, adjusted, and approved by the management. Long-range, 5 year plans are also reviewed to determine what, if any, major capital equipment will need to be procured in order to support the anticipated product demand.

With the sales projected for the year, monthly and weekly numbers are estimated and factored into the overall business plan in order to determine *budgets* for the operating groups within the company structure. Budgets consist of both manpower and operating expenses and serve as a cost target for supporting the projected sales forecast.

Based on the budget, engineering staffing levels and expected expenses such as travel, software, consulting, and supplies are determined for the year. More on that in the next section.

Planning and scheduling are critical activities in the manufacturing organization usually with the objective to provide the most reliable and fastest delivery times in the industry. Some necessary materials will be pulled from current inventory, others will be ordered from suppliers. Inventory-control software is used to determine when additional material must be placed on order. Planning and scheduling personnel issue requisitions to buyers in the purchasing department requesting specific materials and a "need" date. Often material-procurement times dictate when the end product can be shipped.

Manufacturing engineers prepare internal orders for parts to be produced in the shop or laboratory in accordance with engineering drawings. Both internally produced parts and those procured from outside vendors are forwarded to the assembly area, based on the manufacturing schedule. Here, workers/technicians assemble the finished product, which is then moved to the *inspection and test* department. Not only does the quality control department inspect purchased and internally manufactured components but they also verify the assembled product meets the physical performance and engineering specifications. If deficiencies are detected, parts are scrapped or routed for remanufacture. *Test technicians* verify product performance to a known set of

Technical Career Survival Handbook. http://dx.doi.org/10.1016/B978-0-12-809372-6.00044-X

standards before advancing the product to the shipping department. The actual date and time the product ships are tracked against the scheduled ship date to determine the "on-time" shipment performance matrix.

When products reach the buyer's facility and do not perform according to expectations, deficiencies are reported to the manufacturer's *field service* department for corrective action. This may require a field service engineer to visit the product location, trouble-shoot, and make necessary repairs/modifications or order a replacement.

I have spent more than half of my career in the manufacturing world. While it is true that many engineers, scientists, and technicians comprise the manufacturing labor force, there are also a significant number of workers in the shops, laboratories, warehouses, and even in the offices whose specialized knowledge contributes greatly to the overall success and survival of the company.

Song: "Don't Stop Believin'"
As popularized by: Journey
1981

Engineering Department

In previous sections, I discussed specialties, job titles, and the company organization structure. Let us look at the engineering organization in more detail. It all starts with the sales forecast and the resulting shipping schedule, and engineers are usually involved with both.

Before an order is received, the product must be developed and tested. This is usually the job of the *development engineers*, sometimes referred to as research and development (R&D) group. Their output is the future product or product features in prototype form, not ready for prime time. The Sundstrand Fluid Handling Division (SFH) of Sundstrand Corporation came into existence beginning with a prototype that was a high-speed water-injection pump designed and manufactured by the Sundstrand aerospace division for jet fighter planes. From that single prototype, SFH, over many years, designed a family of industrial pumps and compressors for a variety of industries. What other more famous prototype became the basis for a giant industry? Hint: A. G. Bell Fig. 45.1.

The *product engineering* group (aka commercial engineering) must design a production version of the product or a family of products that can be manufactured in quantities anticipated for the marketplace. So in a chicken and egg fashion, the sales forecast must provide a quantity projection such that the production version can be designed according to the anticipated quantities. In order to prepare the forecast, the sales forecast must be based on the predicted product performance, size, and cost. In the manufacturing world, quantities dictate the production methods, that is, cast versus machined and make versus buy.

When a potential sale is identified, a *sales engineer* (aka application engineer) may become involved at the request of the field sales personnel in order to define how the product should be configured and perform based on the customer's specific application. In many cases the customer may submit a written specification defining the product size, performance, and terms and conditions. Often, the manufacturer then prepares product drawings, performance guarantees, and provides the price and delivery schedule and submits their "Proposal" (aka bid). Their bid may be one of many that will be evaluated by the customer before a final award is determined by the buyer.

In the previous section, I mentioned the *quality control* department and its role in ensuring product conformance to established standards. So as to not create a conflict of interest with manufacturing and shipping demands, the quality control function is often under the engineering umbrella or a separate, autonomous organization of engineers, technicians, and inspectors. They may be known as the company cops and can stop or delay shipments depending on their inspection results. For example, they may identify and prevent a common manufacturing syndrome known as "short ship," meaning the product goes out the door incomplete but gets counted as a shipment. The deficiency then has to be dealt with as a field problem and may jeopardize future sales.

Technical Career Survival Handbook. http://dx.doi.org/10.1016/B978-0-12-809372-6.00045-1

Figure 45.1 A family of products derived from a single prototype.

Field *service engineering* is the chief feedback means for the factory to evaluate whether products require "fixes" or modifications to be incorporated to prevent future products with the same deficiencies from being shipped. It is important to nip in the bud any product discrepancies before a negative reputation develops in the marketplace and corrupts future sales efforts.

At SFH, we took a precautionary step to prevent a failure-to-launch situation with new products. We would identify a customer with a need for a pump or compressor who was willing to accept a prototype "field trial" test unit as a solution to a problem application existing in their plant. We would place the prototype for a nominal or below market price with the agreement to fully support the product with parts and on-site service for an extended warranty period. After a number of successful prototypes were placed, we moved forward with a commercial version of the product and incorporated any and all fixes that were made during the trial period. Sales literature was then generated based on the final configuration and performance.

Many highly engineered products are sold either by distributors or retail outlets thereby avoiding customizing the product for a specific application. The best example would be personal computers whereby the configuration is standardized prior to releasing the product to the marketplace.

Song: "I Love This Bar"
As popularized by: Toby Keith
2012

Plant Engineering

46

Chapter Outline

There is a uniqueness about plant engineering function because plant engineering personnel are the behind-the-scenes jack-of-all-trades folks who maintain the operations of power plants, manufacturing establishments, and chemical processing plants. They constitute typically a multidiscipline department/organization consisting of engineers and technicians reporting to a manager who is sometimes referred to as the chief engineer or plant manager. Plant engineers must be flexible to survive.

An engineering friend of mine, Jack Smith (seriously), graduated from Penn State University with a BS degree in nuclear engineering in 1975. Initially he wanted hands on, plant engineering experience prior to joining a big nuclear plant design firm such as Westinghouse or General Electric.

He accepted an entry level job as a plant engineer with Virginia Electric & Power Company, now Dominion, at nuclear plant under construction in Virginia. His plan was to participate in the initial startup of the new reactors to learn all he could in the field and then relocate to a big firm that designed reactors. However, a significant intervening event occurred on March 28, 1979 that would shake the foundation of the nuclear industry for years. A serious nuclear accident occurred during the startup of Three Mile Island (TMI) plant in Pennsylvania. After the TMI accident, Jack was concerned that his career as a nuclear design engineer would not materialize. Orders for new reactors were stalled and many on the drawing board canceled.

Fortunately, Jack was reassigned to lead a team responsible for testing all critical safety systems for the new nuclear unit prior to being licensed for full-power operation. While the plant engineering work alone was challenging, it came with increased regulatory scrutiny and requirements of the Nuclear Regulatory Agency (NRC) as a result of the TMI accident. This unit startup was event-free and very successful becoming the first new nuclear unit in the United States to obtain a license for full-power operation following the TMI accident. He was later assigned to supervise various plant engineering groups responsible for technical programs such as reactor engineering, periodic surveillance testing, design engineering, safety engineering, maintenance engineering, and system engineering. Jack remained with Dominion on that alternate path in various positions until his retirement.

Technical Career Survival Handbook. http://dx.doi.org/10.1016/B978-0-12-809372-6.00046-3

What Are the Duties of Plant Engineers?

Energy consumption, water, cooling and heating systems, compressed air, electric power consumption, and pollution are all important considerations for plant engineering personnel. Also, as part of their daily routine, plant engineers deal with seemingly minor maintenance issues such as compressed air and water leaks, electrical overloads and outages, machinery repairs, and safety issues.

As a typical plant engineering challenge, while I was with ITAC, we assisted Alloy Polymers Inc. (AP) in Richmond resolve an excessive city water consumption problem. As a manufacturer of polymers, water was a critical utility for quenching and setting hot polymer throughout their manufacturing plant. Twelve production "lines" required water baths such that the sum total water flow required was 160 gallons/min. Cool City water was supplied to the plant, routed to various lines, discharged hot from the machines, and piped to the sewer. It was a "one pass" water use system, and consequently water consumption amounted to a major annual operating expense.

AP took the initial step at sizing a cooling tower and placing an order for delivery of the tower in 3 months. At the time, I met with the AP plant engineering personnel to advise how we would configure a "closed-loop" cooling system to conserve water and reduce operating expenses. We settled on installing a water storage tank that would collect water from the cooling tower and then pump the water throughout the plant via the existing "city" water piping. We designed and fabricated several "carts" that included filters, valves, a pump, a tank, and connection hoses. The carts were then wheeled to the various manufacturing lines, connected to the water supply and return, and cooling water was pumped through the machinery. Then the heated water would be returned to the storage tank where it is pumped to the cooling tower for temperature reduction rather than discharged to the city sewer piping.

What Are the Requirements for Plant Engineers?

Many current plant engineers are technicians and hold AA degrees or have technical military backgrounds. However, most applicants today will need a BS degree in engineering of varying disciplines. Different industries have preferences for the specific disciplines they require depending on whether they are a manufacturer, healthcare facility, chemical process, or electrical power generation plant.

What Is the Outlook for Plant Engineers?

The US Bureau of Labor Statistics predicted that plant engineering positions will grow by 5% between 2012 and 2022. This is substantiated as about half of their plant engineering readership are over the age of 55 and their retirement will create a major void in the workforce according to recent research conducted by Consulting-Specifying Engineer magazine. The void will likely be filled due to more hiring and mentoring of technical personnel.

What Are the Trends in Plant Engineering?

Water conservation and energy-efficient buildings will be a cause for concern for building owners, engineers, and municipalities. Legislators will push for water-efficient fixtures, water reuse systems, and high-efficiency boilers. More demand is foreseen for "smart" building technologies that utilize products that can be controlled, monitored, and managed from across the planet. Increased healthcare facility expenditures will likely create opportunities for plant engineers for decades.

Who Is a Certified Plant Engineer?

A Certified Plant Engineer (CPE) is a credential earned through the Association of Facilities Engineering for candidates who possess a combination of work experience and education and who take and successfully pass an 8-hour multiple-choice exam. A CPE accreditation is not a mandatory requirement for plant engineers but it demonstrates competency to perspective employers and may help achieve advancement with existing employers. Core competencies in the CPE program include: electrical, mechanical, environmental, and civil engineering, heating, ventilation and air conditioning, controls, management, economics, maintenance, energy, and Occupational Safety and Health Act (OSHA) regulations.

Song: "Everything is Broken"
As popularized by: Bob Dylan
1989

Service Companies and Consulting Engineers

47

Many engineers, scientists, and technicians are employed by companies that do not provide a product but offer services to design, specify, build, test, and troubleshoot. They are referred to as consulting engineers (CE). While they exist across many different industries, typically they serve only a select group of industries depending on their specialization.

Here are few examples of CEs with their respective services:

Architects and engineering (A&E) firms: Schools, stadiums, airports, and industrial plants.
Mechanical, electrical, and plumbing (MEP): Electrical power, chemical plants, waste treatment, petrochemical, environmental, structural, and fire protection.
Testing: Materials, chemicals, water, systems, and software.
Geotechnical: Roadways, bridges, mining, test boring, core sampling, and exploratory excavations.
Fire protection: Design and install fire-protection systems, sprinklers, controls, and alarm panels.

Having worked for several consulting engineering companies, here are some typical characteristics that I have observed:

Size: Most tend to be small, under 100 employees, and privately owned. They are often founded by an individual consultant for a particular client company initially, forming an organization to provide a single-discipline support to that client, then expanding and offering multidiscipline capability to several clients.
Sales forecast: Manufacturers predict product sales based on feedback from their field sales, distributors, and market data. CEs have a more difficult challenge trying to forecast their business because it must be based on their existing and future clients' forecasted business activity. This data is usually company proprietary and not typically shared. In limited cases, an alliance is formed to enable more accurate forecasting. More on that topic later.
Projects: They often start out on small projects, then bidding and landing larger projects in hopes to be able to recruit the required technical staff in time to support the work. At Adtechs, a CE company, project manager Ed Day hired me in a rush when he found out they were the successful bidders on an effluent treatment facility (ETF) plant in Hanford WA. He recognized that there would be multiple fluid handling systems requiring many types of pumps. I had a heavy background in the design, application, and manufacture of industrial pumps, so I was brought on board early in the project.
Discipline needs: Depending on the projects, often multiple disciplines are required to complete the work. For example, the Hanford project required a team of mechanical, electrical, controls, structural, software, and chemical engineers to complete the design of the ETF

Technical Career Survival Handbook. http://dx.doi.org/10.1016/B978-0-12-809372-6.00047-5

housed in a dedicated building on the government-owned Hanford Reservation. On the other extreme, I have worked on small projects where only the mechanical engineering discipline was required to complete the project.

Organizational structure: In a previous section, I explained the product/project organization, which is commonly used by consulting engineering companies. At the dashboard of a project is the project manager (PM) who assembles a team with the necessary disciplines to accomplish the project scope of work. The PM is also responsible for maintaining the project on budget and schedule. Progress is reported by the PM to the CE company management on a weekly basis and to the client company as well.

Salary structure: Much unlike manufacturing companies who hire, train, and promote employees over many years, CEs tend to think shorter term. Therefore they must employ personnel that are well seasoned in their particular discipline and industry in their employment history. This usually translates to higher salaries for the average employee when compared with those in most manufacturing environments. Often employees are hired promptly as the CE receives a purchase order (PO) from a client company who expects the design work to begin immediately. I was hired and on the job within two weeks in order to start on the Hanford project at which point we were already one month behind the schedule. Years earlier, I was surprised during an interview with a relatively small CE firm in Virginia. They were ready to make me an offer; however, it was contingent upon me making an investment in the company thereby becoming a partner. Since I was not comfortable investing in the company, I declined to pursue the possibility further.

Hiring: Related experience is an important requirement when hiring technical personnel. It is not unusual for a CE firm to hire retirees or laid-off employees from a client company often on a part time, as-needed basis. This provides the CE with employees that are knowledgeable of the client's facility, products, engineering documents, and personnel. Of course, those hires must have departed from the client companies under favorable circumstances. Additionally, it is not unusual for a CE to hire a private consultant temporarily based on a specific technology void in support of a client company.

On the other end of the spectrum, there are a few MEP giants such as Jacobs, Black and Veatch, and Burns and Roe that are billion-dollar corporations. They specialize in new construction and retrofit/renovations for projects such as health care facilities, office buildings, colleges, military facilities, data centers, and many more. Most are privately owned and their sources for work come from architects, design-build contracts, and various privately owned firms.

Song: "All for You"
As popularized by: Sister Hazel
2009

Government

48

Back in the 1960s and 1970s, most engineering graduates overlooked unappealing opportunities in government branches. That was then, but now it is vastly different largely due to government involvement in aerospace, warfare, and environment issues. Government departments hiring technical personnel include Department of Defense, Environmental Protection Agency, Department of the Navy, Department of the Army, Department of Transportation, and many more.

These positions within the government are classified and qualified as Professional Engineering Positions, 0800. According to the Office of Personnel Management (OPM), this group: "Includes all classes of positions, the duties of which are to advise on, administer, supervise, or perform professional, scientific, or technical work concerned with engineering or architectural projects, facilities, structures, systems, processes, equipment, devices, material or methods. Positions in this group require knowledge of the science or art, or both, by which materials, natural resources, and power are made useful."

The 0800 classification has two basic individual occupation requirements as follows:

Degree—obtained from an engineering school with at least one curriculum accredited by the Accreditation Board for Engineering and Technology (ABET) as a professional engineering curriculum. Or alternatively, the curriculum must include differential and integral calculus and courses in at least five studies such as statics, dynamics, strength of materials, fluid mechanics, thermodynamics, electrical fields, properties of materials, and other comparable areas.

Combination of education and experience—a college level education, training, and/or experience that resulted in a knowledge of physical and mathematical sciences plus adequacy of a background in engineering sciences and techniques under one of the engineering disciplines as demonstrated by:

- professional engineering registration;
- satisfactory passing of the Engineer-in-Training (EIT) examination;
- GS-5 level, limited to positions closely related to the discipline;
- specified academic courses totaling to 60 semester hours;
- related curriculum.

Under the 0800 classification, a candidate meeting the above requirements would qualify for a GS-5 level and the corresponding salary range. To qualify for a GS-7 position, candidates would need to achieve superior academic results at the baccalaureate level in a professional engineering curriculum. For a GS-9 position, a combination of academic excellence and at least one year of professional experience is required.

Technical Career Survival Handbook. http://dx.doi.org/10.1016/B978-0-12-809372-6.00048-7

As for salaries, government employees earn an average of almost 60% higher than the *average* private-sector employee, $79K versus $50K, according to the Heritage Foundation in a 2010 USA Today article. The article points out that the Federal workers have more education on the average than private-sector workers.

Excellent benefits include multiple retirement plan choices, health benefits, paid leave, group insurance, and often on-site child care. An additional perk is near-absolute job security. Jobs are not eliminated when sales drop as in the private-sector. It is virtually impossible to fire a Federal employee and it is no surprise that Federal employees rarely quit.

Under the above OPM scope of the duties for 0800 technical personnel, the words "advise on, administer, supervise" summarize the fact that much of the work is associated with construction and maintenance of Federal properties. As such, a majority of the positions will be related accordingly, that is, to civil engineering, electrical engineering, environmental engineering, mechanical engineering, and computer engineering: for example, the Hanford ETF mentioned previously was overseen by government environmental engineers, the site was managed by Westinghouse engineering, and designed and constructed by the general contractor Adtechs, my employer at the time.

Thus, you can see how things have changed in the last 50 years making technical careers in the government sector quite desirable at the taxpayer's expense. Chances for career survival are high. But as Will Rodgers once said, "Be thankful we are not getting all the government we're paying for."

Song: "Summertime Blues"
As popularized by: Eddie Cochran
1957

Union Versus Nonunion

49

Having worked within only two union environments and the rest nonunion, I can comment on the two environments from the perspective of an engineer. That is the intent of this section. However, let us first understand the main differences in union or closed shops versus nonunion or open shops as they are often referred.

Based on a study done by Bankrate, a financial and lending institution, here are some contrasting differences.

Wages: Median weekly income for full time union workers in 2010 was 28% higher than that for nonunion workers according to the Bureau of Labor Statistics.

Benefits: 35% more union workers were entitled to medical benefits than nonunion workers according to the Bureau of Labor Statistics survey of 120 million workers.

Job security: Nonunion workers may be fired "at will"; that is, for no reason. There are of course several exceptions for discrimination, such as race, creed, age, and religion. However, services of union workers can only be terminated for "just cause" and misconduct through a grievance procedure or arbitration process.

Bargaining: Union workers have more power as a group than individual nonunion workers through a process called "collective bargaining" to negotiate wages, benefits, and working conditions with shop management.

Seniority: In the event of a shop layoff situation, union workers are laid off in accordance with the lowest seniority first. This is also a factor in determining promotions.

So what is the impact of a union versus nonunion manufacturing facility or jobsite on the survival of an engineer, scientist, or technician working in these environments? My opinion is as follows.

Flexibility: As an engineer you often render opinions on how to proceed with manufacturing, test, or construction procedures and you want it done yesterday. Dealing with union personnel creates many hoops to jump through to create change. Several forms must be generated, approvals obtained, and discussions held with all involved. Some would say this results in better quality, and in some instances that is certainly the case. But in many situations, it becomes a major roadblock for an engineer trying to meet a project schedule. But as a positive, once the plan was approved, I could be assured that it would be carried out to the letter.

Skills: I was impressed with the skill levels I observed while working on the construction of a power generation plant and a manufacturing plant; both utilized union workers. In observing and talking to many of the union workers, I realized that they were highly experienced in executing difficult welds, material examinations, setting equipment, and generally following established procedures. Most were mature, focused, and traveled considerable distance to locate near the site on a temporary basis.

Song: "Shiftwork"
As popularized by: Kenny Chesney
2007

Technical Career Survival Handbook. http://dx.doi.org/10.1016/B978-0-12-809372-6.00049-9

Utilities

50

Frequently the news media brings us stories about utilities but generally does not address water and wastewater piping that tends to be a local issue. Although occasionally, a water main break may make the headlines. Contrarily, electricity is always front-page news because of its association with the environment whether produced by a fossil or nuclear power plant.

My first involvement with electric power plants was as a contractor working on the La Paloma Generating Company, LLC, 1124 MW, natural gas-fired, combined cycle facility; see Fig. 3.1. The plant is located near McKittrick, in Kern County, California, about 40 miles west of Bakersfield, California. The combined cycle power block consists of four combustion turbine generators (CTGs), four heat recovery steam generators (HRSGs) and exhaust stacks, and two steam turbines.

My impression of the location when I first arrived on site was that it was a perfect place for a power plant. At 40 miles from a metropolitan area in the desert with nothing but snakes and spiders, what is the problem? There were essentially four permitting problems:

- Construction of a 15-mile bundled 230 kV double circuit overhead transmission line.
- Construction of a natural gas pipeline jointly owned by the Kern River and Mojave Pipeline Co.
- Construction of a potable water pipeline from McKittrick.
- Gas turbine exhaust stack emissions.

I was not involved in the permitting process, but learned that it was exacerbated by the fact that a new power plant had not been constructed in California since 1990. But residents of Los Angeles had been experiencing gray outs and blackouts over the past few years and consequently, there was tremendous pressure on the generation company to take action. While having dinner in an LA sports bar after one of my plant trips, word got out about my work on the power plant, someone yelled it out, and spontaneous clapping and cheering broke out in the bar. While engineers are not often in the spotlight, it was a memorable moment in my career.

Many different trades, engineering disciplines, and technicians were required to complete the construction project which made my first of three positions in power generation challenging. My role with the engineering company, Alstom Power, was to assemble the bid documents and facilitate the selection of the mechanical contractor, a.k.a. the low bidder. The successful bidder's price was actually too low in our opinion and we were worried that they would go out of business mid-way through the construction. So we worked with them over several weeks to better familiarize them with the scope of work. Consequently, the original $30 MM bid rose to what we thought was realistic but not too profitable for them and we cut the purchase order.

Technical Career Survival Handbook. http://dx.doi.org/10.1016/B978-0-12-809372-6.00050-5

Figure 50.1 Power plant owner's construction supervisory team.

My second position was as a contractor to Dominion Virginia Power as part of a site management/project team that I referred to as the "dirty dozen" (see Fig. 50.1). The project was similar to the La Paloma project except that the power-generation plant was an addition to an existing coal-fired power plant site along the Potomac River in northern VA.

After the completion of the Dominion plant, I was hired as an employee in their nuclear division in Richmond VA as part of the mechanical design team managed by Mike Henig. While no *new* nuclear plants were on the horizon, our work was to provide design support for equipment within Dominion's fleet of aging nuclear plants. I learned that a characteristic of technical personnel in the nuclear industry was very low turnover and hence the average employee is past their career midpoint largely due to the inactivity in the nuclear industry.

Similar to fossil power plants, the main objective in nuclear plant is to "keep the breaker closed." By that, I mean action is taken to ensure that the electrical power transmission is continuous, the electrical breaker is in the closed position, and there is no downtime except for annual planned maintenance outages. It follows that safety and maintenance rule the plant, and our design efforts were seen through that prism. Policies and procedures become the focus for all activities, which was a new experience for me. I tried to adapt quickly in order to survive. Design changes came about painfully slow after many reviews, revisions, and approvals.

Song: "Evil Woman"
As popularized by: Electric Light Orchestra
1975

Spinoffs, Acquisitions, and Mergers

51

Chapter Outline

Engineers often feel powerless in the face of corporate decisions that impact their employment longevity and workplace location. But the truth is, nothing is more constant than change. So it goes with *spinoffs*, when business units are separated, *acquisitions*, when business units are purchased, and *mergers*, when business units are combined. Often during these transactions such as spinoffs, for example, it is thought that it is a positive for one unit and a negative for the other. The consequences of these business decisions can be very disruptive for technical career personnel as they often result in downsizing, reorganizations, or relocations. While I did not directly experience any of these during the course of my career, looking back, I find that one of my previous employers demonstrated an excellent example of spinoff and acquisition phenomena.

Spinoffs

In 1957 Sundstrand Corporation successfully designed, manufactured, and shipped a gear-driven, light-weight, high-speed water injection pump to a commercial jet aircraft manufacturer. The pump increased the engine thrust when high-pressure water was injected into the combustion chamber. The subsequent demand grew and Sundstrand had a profitable product to add to their aircraft components' product line.

Based on the aircraft version of the high-speed pump, Sundstrand derived a heavy duty industrial version of the product and sold a prototype to Shell Chemical in 1962. The trade name for the product became known as the "Sundyne pump." Again they achieved success with the derivative product. However, Sundstrand Corp. was in the aerospace business and did not have an established sales network to reach industrial markets. So an industrial product business unit, Sundstrand Fluid Handling (SFH) Division of Sundstrand Corp. was created as a spinoff in Denver Co. initially sharing a manufacturing facility with the sister aerospace division. Products were sold through a network of direct sales personnel, manufacturer's representatives, and distributors.

Technical Career Survival Handbook. http://dx.doi.org/10.1016/B978-0-12-809372-6.00051-7

Later, I was recruited as the manager of development engineering in 1976 in time to experience the move to a new and separate manufacturing and test facility in Arvada Co. My role was to continue the successful development of high-speed pumps, compressors, and blowers for a variety of industrial markets. Business was brisk and many young engineers and technicians were hired who enjoyed the attractiveness of the Colorado Rockies.

Acquisitions

SFH continued to prosper under the control of Sundstrand Corporation. As they grew and expanded their sales network, they also acquired several pump companies and formed a joint venture with a Japanese corporation; all the while determined to acquire compatible products and avoid overlapping capabilities.

Then in 1999 came the big surprise. United Technology Corporation (UTC) successfully bid to acquire Sundstrand Corp. and merge it with Hamilton Standard Division of UTC (my first employer after graduating) under a new name Hamilton–Sundstrand. However, since SFH was a nonaerospace division, it became the Sundyne Corporation under Hamilton Sundstrand Industrial Division subsidiary of UTC. Pretty straightforward, right? The story does not end there.

In 2012 Sundyne Corp. and Industrial Division sister companies Sullair and Milton Roy were spun off by UTC to a joint venture owned by the Carlyle Group and BC Partners to form what is now named Accudyne Industries.

Mergers

Probably the least common of the three phenomena but the most disruptive is the merger, where business units are combined. The difficulty lies in having sufficient commonality in the business units to realize a significant benefit in the union. If there is duplicity in the units to be merged, then the consequence of the merger is usually eliminating personnel and forming a leaner organization with reduced overhead. While favorable for the company, this is not favorable for certain employees.

Some would say spinoffs, acquisitions, and mergers are extremely disruptive in the life of an engineer, scientist, or technician. Well, yes in a way, but if it keeps the company solvent and profitable, that is a big plus. In fact, I landed a product manager position at Worthington Pump Div. in Maryland as a result of the Studebaker–Worthington Corp. in N.J. spinning off two product lines—engineered pumps to Maryland and standard vertical pumps to Oklahoma. I considered Maryland desirable as it was close to my future wife's home.

To illustrate how frequently spinoffs, acquisitions, and mergers can occur, I went back in time in order to see if I could have survived if I had stayed with the Studebaker–Worthington Pump Corp. What I found was:

1985: Dresser Industries acquires Studebaker–Worthington Pump Corp.

1992: Ingersoll Rand Corp. merges with Dresser pumps
2000: Flowserve acquires Ingersoll–Dresser pumps

Song: "I Will Survive"
As popularized by: Gloria Gaynor
1978

Startups

52

I mentioned previously that Adtechs Corp., a Japanese-owned CE company, hired me in a rush when they found out they were the successful bidders on an Effluent Treatment Facility (ETF) plant in Hanford, Washington. They were well established in Asian countries based on their engineering design and construction of hazardous and radioactive treatment facilities. So they made a business decision to startup an engineering office initially in Richmond, Virginia, but shortly thereafter relocated to the Dulles Airport area for convenience to the Environmental Protection Agency and Superfund officials.

I recognized the risk in signing up with an unknown company in the United States, but was willing to take a gamble that they would be equally successful in the United States as they were in Japan. Plus, Japanese companies had a reputation in the 1980s for being in business for the long haul. Similarly, they recognized there would be multiple fluid handling systems requiring many types of pumps and were willing to take a chance on hiring me as I had a heavy background in the design, application, and manufacture of industrial pumps.

Based on my experience, I think the best word to describe this particular startup situation is "chaos." Our first office in Richmond was cramped and every available square inch was utilized for desks, copy machines, and computers. It was initially hard to understand the organization structure and it seemed to change on a daily basis with the arrival of technical personnel from Japan and those hired from within the United States who just seemed to appear unannounced without any notice. Not to mention that verbal communication difficulties alone were unending, particularly with my boss "Bart" (not his true Japanese name of course).

"Exciting" is another word to describe the work. I did not actually have my work planned out, I simply worked on what I was told was the emergency of the day and hoped I was making significant progress. Then came the relocation to the Dulles area, which was somewhat anticipated when I was hired. They told me that they would provide financial assistance if I moved my family near the office, but after spending a day with a realtor, we nixed the idea. I decided to try commuting daily, 80 min each way. That is not unusual in the D.C. area.

We expanded our employee numbers at a rapid clip, and finally after a year into the business we had a well-defined technical staff. Because of the newness of the industry we were serving and our geographic location, hired personnel came from a very eclectic variety of industries such as missile manufacturing, government, power, engineering and construction, machine manufacturing, and design. This mix made for great conversations during after-work happy hours.

Despite assembling a technically astute organization, the company experienced difficulties managing the government-funded ETF project at Hanford. You might say we

Technical Career Survival Handbook. http://dx.doi.org/10.1016/B978-0-12-809372-6.00052-9

were bullied into complying with numerous design change requests that were passed down. Many of the changes affected equipment that was being provided by manufacturers and suppliers who passed cost increases on to Adtechs who was contractually unsuccessful passing them on to the client. After about one year into the project, it became apparent we would lose a substantial amount of money at the close of the project. Although there was always that prevailing attitude that the Japanese had deep pockets and were in it for the long haul.

Finally after four years of red ink, a gradual decline in the company financial health, and many hours sitting in traffic jams, I elected to resign after accepting a position with a Richmond based consulting engineering company. I had no regrets about my tenure with Adtechs and was thankful for the excitement that accompanied the experience. My experience with Adtechs allowed me to transition from the manufacturing industry into the consulting engineering world, surviving without a negative reboot and fortunately without disruption for my family.

Song: "It Feels Like the First Time"
As popularized by: Foreigner
1977

Alliances

53

A single, mutually beneficial legal agreement between two or more businesses with equity and opportunity risk for all parties is an alliance. Usually the motivation behind the agreement is for controlling cost and/or enhancing service. I had experience with this business arrangement strictly from a *sales perspective* while serving as the marketing manager for Kaeser Compressors. Their products were utilized primarily in plant compressed air systems applications. These systems required air dryers and filters to remove particulate and moisture from the delivered air. Hankison Corp., headquartered in Pennsylvania, manufactured dryers and filters for compressed air.

In the late 1980s, Kaeser and Hankison formed an alliance such that the latter would supply the former with air dryers and filters under the name and color scheme of Kaeser. Additionally, literature for the product lines was prepared by Kaeser based on Hankison performance data. Brochures were forwarded to distributors for the sales purposes. The alliance benefited both parties. First, it allowed Kaeser to provide a turnkey system of air handling components from a single source. Second, it resulted in additional product sales for Hankison without expanding their sales force.

In another situation, I experienced an alliance from a manufacturing standpoint. Because process industries are constantly upgrading equipment, improving productivity, and expanding their capacity, they often require year round engineering support. Largely due to my recent experience with Adtechs, I was hired by Day & Zimmerman consulting engineers shortly after they signed an alliance agreement with chemical manufacturing company Honeywell Corp. in Hopewell, Virginia, which had three plants in the Richmond, Virginia area. The alliance required D&Z to ramp up their staffing in short order to be in a position to provide engineering manpower to support Honeywell's business plan. Within a few weeks, we had technical staff literally working out of closets and conference rooms.

An alliance between an engineering company and a manufacturer is typically established as follows:

1. The engineering company determines rates based on W-2s for their technical staff based on a 40h week without overtime. They tack on 4–8% for overhead, such as HR, reproduction, and accounting staff, and profit; then provide that as the billable rate to the client company. Overtime rates are determined similarly.
2. As the engineering company encounters expenses such as travel, reproduction, and permit fees while engaged in the client company projects, those costs are passed on without markup.
3. Alliances are contracted for a specific time period like 3 years, cancelable 90 days in advance and optionally extended for another year or two.
4. In order for the engineering company to make allowances for peaks and drops in the work load, they may need to supplement their work force with temporary technical staff. Also they

Technical Career Survival Handbook. http://dx.doi.org/10.1016/B978-0-12-809372-6.00053-0

may need to hire replacement personnel as people leave through attrition, particularly those in high demand.

5. The client company may provide feedback to the engineering company with regards to employee performance for salary and benefit adjustment purposes.

An alliance is intended to be a win–win situation for both parties. It allows each party to focus on their main business. However it requires trust, communication, and vision to be successful. Technical staff involved in the alliance must be cautious about the security of their participation in the agreement and not be caught off guard. Employees for the engineering company must always be aware of the agreement duration because contract renewal is not a sure thing and definitely affects survivability, I discovered.

Song: "Happy Together"
As popularized by: Turtles
1967

Joint Venture

54

In 1970 Sundstrand Fluid Handling Division (SFH) formed a joint venture with Nikkiso Company in Japan. A *joint venture* (JV) is a mutually beneficial business agreement in which the parties agree to develop, for a finite time, a new diverse business unit with new assets by contributing equity. The JV is separated legally from the two parties, but both the parties have control over it. SFH products were shipped to Japan and sold by Nikkiso–Sundstrand.

In 1984, while I was manager of development engineering for SFH, a disaster occurred at a Union Carbide chemical plant in Bhopal, India. Certain equipment within the pesticide plant leaked methyl isocyanate gas and other chemicals creating a dense toxic cloud over the region killing more than 8000 people within just a few days.

Although the exact cause of the leakage was not determined, component and equipment manufacturers reacted to this disaster by improving the reliability of their equipment sealing configurations. In response, SFH under the JV agreement began to market "Seal less pumps" aka canned motor pumps (CMP) *manufactured by* Nikkiso in Japan. The CMP pump configuration contained an electric motor housed within the pump casing and utilized the internal pumping fluid to lubricate the rotor shaft bearings. Hence there was no rotor seal to leak fluid overboard. The CMP became a standard in the chemical industry over the next decade.

The SFH–Nikkiso JV became very successful within just a few years. So much so, that SFH began to manufacture CMPs under the JV at their manufacturing facility in Arvada Co. The company reorganized as a result, which I will explain in a later section. Then the JV: communications between parties or "co-ventures" became frequent—both written and face to face. Both parties invested equally under the agreement with respect to monetary arrangements, technological expertise, and technical service.

Therefore, other than doing nothing, you might ask what were the alternatives to the SFH–Nikkiso JV? Answer: A very ambitious plan for both parties. SFH would need to manufacture a CMP product line using Nikkiso drawings, gain field experience through trial units, and then proceed with production. The other party, Nikkiso, would need to establish and train a sales organization and warehouse products for shipment within the United States, a difficult route for both parties.

Other more famous JVs in the past include Dow Corning, MillerCoors, Penske Truck Leasing, Isuzu and General Motors, and Owens–Corning. Most would call these a win–win. My experience with the JV was certainly positive primarily because of the cooperation on both sides of the agreement.

Song: "Just the Two of Us"
As popularized by: Bill Withers
1981

Technical Career Survival Handbook. http://dx.doi.org/10.1016/B978-0-12-809372-6.00054-2

Job Site Work

55

Most technical personnel will experience the office job environment probably from day one of their career. There is a lot to be said about the conveniences of working in an office: comfortable temperatures, relatively quiet atmosphere, conference rooms, break rooms, well-lighted bathroom facilities, and maybe even a cafeteria. This was my world for most of my career but occasionally I would venture into the field for a brief meeting or to examine a piping or equipment problem. Then came my on-site job as an engineering contractor for a Dominion Virginia Power project in 2003. I bring this experience up as I believe it is a great example of how construction projects proceed and what technical personnel might experience.

The project consisted of the construction of the combined-cycle gas turbine power plant number 6. The combined-cycle unit at Possum Point, Virginia on the Potomac River includes two combustion turbines and a steam generator. The combustion turbines are similar to jet engines. Forced air is superheated by natural gas or oil that turns a gas turbine to generate electricity. Exhaust from the combustion turbines is directed to a heat exchanger and used to produce steam to power a steam turbine. By utilizing the waste heat to produce power, the combined-cycle units burn less fuel and produce fewer emissions than comparably sized conventional units.

I was part of the owner's construction supervisory team consisting of mainly contractors like me and Dominion employees tasked with managing the engineering and construction of the plant and representing the owner's interest. The team was led by the project manager Emil Avram, a Rensselaer Polytechnic Institute graduate and mechanical engineer who established the team in a line organization (see Chapter 50). Because the team members were highly experienced in their disciplines, the organization was broad and not layered, see Fig. 50.1. Occasionally, specialized personnel from the Dominion corporate office joined the team for a few days to resolve a design issue. The following disciplines and personnel constituted the team:

Mechanical engineering
Civil engineering
Electrical engineering
Instrument technician
Schedule coordinator
Document clerk
Quality control inspector
Safety technician

Although the overall schedule was important, the emphasis on safety was profound. During the construction, there are many temporary provisions to enable workers to install the permanent aspects of a power plant and most other facilities. As a result, accidents can occur if special precautions are not observed. According to the United

Technical Career Survival Handbook. http://dx.doi.org/10.1016/B978-0-12-809372-6.00055-4

States Department of Labor, 6.5 million people work at approximately 252,000 construction sites on a daily basis. They face major hazards, such as falls, trench collapse, scaffold collapse, electric shock, and failure to use personnel protective equipment (PPE).

Therefore, there are risks to workers that need to be emphasized through worker training, daily meetings, and signage. Every morning meeting began with a safety topic and discussion about the previous day safety observations. Topics were often derived from the ten most frequently issued citations of Occupational Safety and Health Administration (OSHA):

1. Scaffolding
2. Fall protection
3. Excavations, general
4. Ladders
5. Head protection
6. Excavations, protection
7. Hazards communication
8. Fall protection training
9. General construction safety
10. Electrical design and protection

The job site was located on a "brown field," meaning it was formerly used as a staging and storage site, while other portions of the power plant were under construction dating back as far as 1950. Hence, considerable emphasis was placed on restoring the site to a "green field" environment while simultaneously constructing unit number 6. Excavation work was frequently halted when an old 55 gal drum or vessel was unearthed causing a Hazmat team to examine the condition of the soil and rendering an action plan. All action and findings were well documented before work resumed.

Surprisingly, I enjoyed the on-site job experience and found it challenging. However there was one major negative, the coldest winter along the Potomac in a decade. So cold in fact, the river froze stranding cargo ships in the channel for days. I have the utmost respect for those power plant contractor workers who were exposed to the elements up to 12 h per day although the project team spent the majority of our time in relatively comfortable temporary trailers except while resolving problems in the field. My work dealing with contractors and mechanical issues came to an end as plans for commissioning the plant proceeded in preparation for plant turnover to the client/owner. While I was worried about survivability, I had no idea that within a few months, I would be hired as a direct Dominion employee in their nuclear division.

Song: "Born in the USA"
As popularized by: Bruce Springsteen
1984

Commuting/Telecommuting

56

In the past, we chose how far away from the office we were willing to live. We traded convenience for desire to live in suburbs or the country and the cost of commuting to work versus cost of housing. Generally if we worked for a manufacturer, it was undesirable to be in walking distance to work, hence we commuted. That was the case until the advent of computers and the Internet. Then for many workers it became feasible to work from home, thus the term "telecommuting" arrived. On the surface, it appeared that this was the solution for reducing expenses associated with operating a vehicle, reducing pollution, and spending hours on the road in traffic.

I have always placed a high priority on where I live perhaps even higher than where I worked. Maybe that is why for many years I had to tolerate commuting well over an hour each way to work. This allowed me to work for many companies without relocating and disrupting my family's life. While working in the Dulles area outside D.C., I chose to commute rather than to move close to the office in a high cost-of-living area. While this was not perceptive on my part, in retrospect, I realized that it is wise to live in a strategic location.

All of these factors may seem of little relevance to a technical career in particular. Therefore, let us look at this subject in more detail. The advantages of telecommuting have been mentioned earlier and they appear fairly obvious, hence let us consider some of the negatives associated with telecommuting in conjunction with technical careers:

1. Many engineers work closely with designers and technicians utilizing specialized software. Often the software is provided on a per seat cost basis and not easily accessed remotely. Plus, working as a team with designers is less effective if one member is located at home and the other is in the office. If a technician is conducting a test, it behooves the scientist, engineer, or designer to maintain frequent face-to-face contact throughout the process.
2. While much of the technical references may be accessible remotely, many may be company proprietary information and not readily accessible from home. Much of the technical information may not be in electronic format and stored in secured file cabinets in the office.
3. Often impromptu interdiscipline meetings may be held to discuss design problems, or equipment/ component vendors may visit unannounced to provide procurement progress or product information. These activities would not happen while working at home.
4. Tracking progress of a team member working from home is hindered. This problem was highlighted by the Washington Post in conjunction with patent examiners at the US Patent and Trademark Office. It was discovered that they were not putting in as many hours as they had represented on their time sheets. Additionally, they would postpone the work until late in the quarter, a practice called "end-loading."

Technical Career Survival Handbook. http://dx.doi.org/10.1016/B978-0-12-809372-6.00056-6

From a personnel standpoint, I prefer to work in the office for focus, fewer distractions, the available IT support, and interfacing with coworkers. Having commuted and telecommuted, I found it easier to survive working in the office because (1) too many distractions were present in my home and (2) there was always something I needed in the office.

Song: "Drive My Car"
As popularized by: The Beatles
1965

Reorganizations

57

The word "reorganization" sounds scary for any technical employee in a company. Likewise, the words "restructure," "redefining," or "realignment" are also frightening. Often it is preceded by rumors. In my case, it all started during a management meeting when our General Manager (GM), Bud Wallin, at SFH handed out copies of the book *In Search of Excellence* authored by Tom Peters and Robert H. Waterman in 1984. This was one of many top-selling business management books that were all the rage in the 1980s. The GM went on to explain how inspiring the book was and that he was going to reorganize the company based on some of the principles described in the book: principles, such as "staying close to the customer," "stick to the knitting," and "simple form–lean staff." He definitely got our attention in that early Monday morning meeting.

Here is the background on why he made this wise executive decision.

> At the time, SFH, under the JV agreement, began to market "Seal-less pumps" aka canned motor pumps (CMP) manufactured by Nikkiso in Japan. This would impact the engineering work load in Arvada as the Nikkiso pumps would be received, modified, tested, and shipped to distributors in the United States. SFH also had a general industrial pump line that was becoming a significant profit generator. These pumps were sold through stocking distributors throughout the United States. Then there was a third, mature, heavy duty American Petroleum Institute (API) target market pump line that generated the majority of the profit for the division, essentially our bread and butter. These pumps were sold directly or through manufacturer's representatives. SFH was organized as a classic line organization (see Fig. 34.2). Therefore the GM's main concern was our ability to focus on all three product families regardless of their revenue streams, product familiarity, and markets.

Within a week from that Monday morning meeting, the GM announced his plan. SFH would be reorganized into three product groups: API, distributor, and CMP lines. Each group would fall under a product director and have its own engineering, quality, service, marketing, and sales departments. Support functions, such as accounting, manufacturing, human resources, development engineering, and IT would be common to all three product groups. There would not be any additional hiring required for office personnel, but there would be some shuffling in order to provide sufficient emphasis on each of the three product lines. The field sales organization would be changed significantly and necessitate relocation of several members. Instead of one sales manager, the reorganization would require three.

The shuffle was completed within few weeks, and it was not long before a feeling of competition among product groups developed. The CMP group became the most spirited group as they were mildly rebellious new kids on the block—nothing to lose, everything to gain so to speak. I was promoted as director of the API product line. Little did I realize that in a friendly competitive way, my department was harassed as being the "gold brickers" and the big kid on the block. It was a bit strange at first to think in terms of three different groups of personnel in the division. We had morphed into a combination product organization and line organization, not exactly text book style. But equal representation and focus resulted from the reorganization and we began to see a significant improvement in the bottom line fairly quickly.

In retrospect, it was a bold move when compared with many companies within our industry that maintained well-defined line organizations for many years. Not that change is always good, but it is wise to periodically step back and determine if the company is organized for survival and optimum performance. Perhaps it is best to have an outsider to make that assessment.

Song: "Changes"
As popularized by: David Bowie
1972

Part III

The Job Search

Resumes

58

During my brief career departure to my position as a technical recruiter (OK, head-hunter), I saw more resumes than I care to remember in various styles, shapes, and formats. I came to conclude that I preferred a particular format, which I will describe later. However, believe me there is no magic here and whatever works is the best approach.

I recommend you think about the format from the perspective of the personnel who will be reviewing the resume. Generally, when a resume is submitted to a company, the first person to review the resume will likely not be familiar with the technical aspects of the position. A human resource department worker will be assigned *to screen the resumes* and only pass on those to the hiring authority that are relevant. The hiring authority is the supervisor or manager that makes the final decision to hire (sometimes fire) candidates assuming there is sufficient monies in their budget to cover salary and benefits. The HR person is likely familiar with the preferred degree, salary require-ments, and industries the ideal candidates should represent. However, they will not be able to distinguish how well the candidate's experiences and accomplishments relate to the open position. So you must avoid being screened out by HR.

I have found the straightforward approach works best. That is, to provide the sta-tistics on *page 1* and candidate's accomplishments on *page 2* of the resume. Page 1 of the suggested format should list:

Name
Physical address and phone number
Email address
Education
Licenses
Special training or skills
Experience, companies, titles, and dates of employment
Publications, patents, or papers
Personal interests/activities

Earlier in your career, it will be challenging to fill the entire page and that is under-stood. You can list your high school, service groups, scouts, part-time jobs after school, leadership roles, and any special honors you received. As you gain more experience and build your resume, you may find it difficult to squeeze everything on one page. Therefore, I would suggest you delete the older, precollege information so you can *maintain a single page* of personal data. Do not get carried away with providing com-pany addresses, websites, and phone numbers. You will likely be required to complete a company application form with spaces to fill in those specifics. So be certain you keep that data handy in case you need to provide it during your first interview Fig. 58.1.

Technical Career Survival Handbook. http://dx.doi.org/10.1016/B978-0-12-809372-6.00058-X

PETER Y. BURKE, P.E.

8336 Kintail Dr.
Chesterfield, VA 23838

EDUCATION

Rensselaer Polytechnic Institute, MS Mechanical Engineering
Virginia Polytechnic Institute and State University, BS Mechanical Engineering

SPECIAL TRAINING, PARTIAL LIST

Dale Carnegie Human Relations, Patent Law for Engineers, Finance for Non-Financial
Executives, Interaction Management, Kepner-Trego Analysis, B31.1 Piping Analysis,
CAESAR II piping analysis, NSPE-Project Management, OSHA 10 Hours

LICENSES

Professional Engineer: State of Virginia, No. 20867 active, Maryland and New York
licenses inactive

EXPERIENCE

FlowSystems Engineering - Chesterfield VA
    Consultant (part time) 5/07 to present
Dominion VA Power Nuclear-Richmond and North Anna, VA
    Nuclear Engineer III- 5/03-5/07
ITAC (consulting engineers)-Richmond VA
    Consultant-Richmond VA-1/03-5/03
Dominion VA Power/Spear Group (consulting engineers) - Norcross GA
    Project Engineer- 4/02 to 1/03
Day & Zimmermann, Int. (consulting engineers)- Richmond VA
    Lead Mechanical Engineer- 11/96 to 4/02
ADTECHS Corp. Div. JGC Corp. (consulting engineers) - Richmond and Herndon VA
    Consulting Mechanical Engineer-1/92 to 11/96
Kaeser Compressors, (manufacturer) –Fredericksburg, VA
    Engineering Manager, Marketing Manager - 3/88 to 1/92
Sundyne Pump Corporation Div. UTC (manufacturer) - Arvada, CO
    Manager of Engineering, Director of Engineering- 8/76 to 4/87
Worthington Pump Corporation (manufacturer)-Taneytown, MD
    Product Engineer- 4/74 to 8/76

PUBLICATIONS, PARTIAL LIST

"Intercoolers/Aftercoolers" Chemical Engineering, 9/82
"High Speed Pumps" Industrial Energy Conference, 6/86
"Recovering Heat from Air Compressors" Plant Engineering, 3/89
"Avoiding Mistakes in Air Compressor Selection" Plant Engineering, 1/90
"Avoiding Compressor Maintenance Problems" Maintenance Engineering, 4/91
"10 Steps You Should Take to Cost Savings" Plant Services, 6/93
"Haz/Rad/Wastewater Pumping Solutions" Pumps & Systems, 10/95

Figure 58.1 Example resume page 1.

Page 2 of your resume should pertain to your experience and accomplishments.
But here is where you will need to do some customizing. Build an e-file with a sentence or two describing your accomplishments during each of your employment positions. Your accomplishments need not be earth-shaking feats, but should represent

Peter Y. Burke, P.E.
Partial list of project accomplishments, 1/92 to present

• Lead mechanical engineer- specified chemical addition equipment for cooling tower at Dominion/North Anna Power Station, prepared general arrangement concept, developed specifications, selected vendors for double wall PVDF B31.3 piping with heat trace, leak detection and insulation, chemical pumping systems, eye wash station and fabricated skids. Supervised installation and resolved construction problems.

• Project engineer-mechanical, for Dominion Energy at the Possum Point Combine cycle power plant construction site in Dumfries Virginia. Made piping material recommendations, approved design changes and field modifications, prepared vendor RFQs, evaluated contractor bids and change orders, and was responsible for ensuring Dominion Energy quality standards for the project were met.

• Lead mechanical engineer for Alstom Power LaPaloma Project, CA, 1050MW combined cycle power plant. Prepared bid specifications for mechanical contractor selection and evaluated bids. Specified equipment, piping and valves, prepared requisitions and performed factory inspections. Evaluated and negotiated change orders and recommended for approval/revision. Directed A/E firm design changes, estimated cost impact and implemented via Field Service Instructions. Working from both office and site, served as liason with site, office and factory to resolve problems with pipe routing, valves, heat tracing, materials, interferences, testing and equipment installation during construction and commissioning.

• Lead engineer for AlliedSignal Chesterfield VA Plant on an 85 ft X 5 ft diameter upgraded replacement Leacher vessel, pumps and automatic backwash particulate filter in a nylon fiber production process. Prepared equipment specifications, selected vendors, inspected equipment and assisted on site during the field installation. Received a "Success Award" for saving the client $20K by modifying a surplus static mixer thereby avoiding a new purchase.

• Lead mechanical engineer for a Philip Morris 4,000 CFM carbon dust collection system installation in the Manufacturing Center Richmond, VA. Sized equipment, prepared specifications, evaluated equipment vendor bids, prepared installation cost estimates, designed piping, P.E. stamped design documents and inspected equipment prior to shipment to site.

• Lead engineer for AlliedSignal Hopewell, VA Plant Ammonia Stripper Column project. Sized equipment and piping, prepared specifications, prepared requisitions, work orders, P.E. stamped design documents, inspected pumping equipment and provided consulting services during construction. Received a cash bonus award for on-time performance for the engineering and construction.

• Prepared specifications for pumping equipment and supervised the construction of the equipment bases, ASME B31.3 Code metallic and non-metallic piping and instrumentation for a 240 gpm radioactive water treatment plant for Savannah River Site, SC. Worked at the fabricator's facility in Orangeburg SC to ensure proper fabrication, assembly and test procedures were followed.

• Developed P&IDs from flow diagrams, prepared specifications, evaluated vendor bids, analyzed pumps/piping, and inspected equipment. Supervised the fabrication of pump and equipment skids in Toronto Canada for the Ontario Hydro Bruce Station nuclear power plant waste treatment process. Work was completed on time and on budget.

• Sized, specified and selected $7M fluid handling equipment for a state-of the-art Hanford Washington DOE Superfund haz/rad waste treatment facility which included filtration, forced circulation evaporation, thin-film drying, UV Oxidation and a drum conveying system. Stamped P&IDs and PFDs, provided consulting services for fluid pumping and operation problems during construction and commissioning phases.

Figure 58.2 Example resume page 2.

specific types of tasks you dealt with during your past history of employment. For example:

- Lead engineer for AlliedSignal Chesterfield, Virginia Plant on an 85 ft × 5 ft diameter upgraded replacement leacher vessel, pumps, and automatic backwash particulate filter in a nylon fiber production process. Prepared equipment specifications, selected vendors, inspected equipment, and assisted on site during the field installation. Received a "Success Award" for saving the client $20K by modifying a surplus static mixer thereby avoiding a new purchase.

Now here is the key. Before you prepare page 2, review the requirements of the position you are applying for. Then, depending on that position, cut and paste the relevant accomplishments from the e-file onto page 2 of your resume in order to tailor it to the specific job requirements. These accomplishments should be listed in order of most recent to oldest Fig. 58.2.

Here are few additional tips for resume preparation:

1. Make you resume easy to read by selecting a 10- or 12-point font.
2. Do not use common overused phrases, such as highly motivated, results-oriented, team player, and problem solver.
3. Do not list basic skills, such as MS Word, Excel, and Office.
4. Do not list your degree dates.
5. Avoid the repeated use of the word "I."
6. Do not have a cluttered-looking resume.
7. Do not list references. The employer will likely request them later.

Call me old fashioned, but I still like attaching a brief cover letter to a resume. The letter or email need only explain why you are submitting the resume. For example, "as you requested," so and so "thought you would like to see my attached resume," "we spoke briefly about your open position for a …," "I am interested in your open position for a …," and "in response to your recent advertisement in the ….."

Many companies, such as General Electric, provide e-resume forms that require largely fill in the blanks and paragraph insertions. Conveniently, they can be forwarded to those involved in the company decision to invite an applicant to an interview.

Caution: Always bring hard copies of your resume to interviews because people lose papers.

Song: "Get a Job"
As popularized by: Silhouettes
1958

Headhunters

59

After being laid off through a corporate downsizing due to the turndown in the oil and gas industry, I took a leap of faith. Thanks to a generous severance package resulting from 10 years of service, I made a bold decision to become a *technical recruiter* a.k.a. headhunter or executive searcher. I rationalized that I had been recruited, I had used recruiters, and therefore I knew everything there was to know about technical recruiting—wrong.

What appealed to me most about the decision was the portability of the job. You could virtually work anywhere as long as you had a phone. But I decided to take a conservative route and purchased a franchise from Management Recruiters Inc. (MRI) headquartered in Cleveland, Ohio. When I contacted them I learned that the franchises were allocated by counties. Being interested in returning to the D.C. area, they had an opening available for me in Loudoun County, just west of Fairfax County in the D.C. metro area, home of the Dulles airport. Hence, I purchased the franchise priced according to the county population, leased an office, and opened it for business.

A technical recruiter places candidates with employers and in return receives a fee on a "contingency" basis amounting to anywhere from 20% to 30% of the first year's salary. In other words, no placement, no fee collected. Most of the selection process is handled over the phone. Searches are usually confined to senior engineer, manager, and mid-level manager positions. This differs from an *executive search firm* that charges a fixed fee for their producing candidates. This is usually the approach used for very high-level positions, such as company presidents, CEOs, and college presidents, where several candidates are carefully scrutinized before a decision is reached. Often a board or search committee will render the hiring decision.

Basically the way they conduct searches is to start day one with a candidate's resume and market the candidate to companies that may possibly have an opening. They contact a possible hiring authority (HA) and present the candidate's qualifications over the phone. If the HA expresses an interest in the candidate and agrees to pay a fee if hired, the recruiter then sets up a future telephonic interview between the HA and the candidate. Further, during that initial call, the recruiter asks if there are other open positions that candidates could fill, thus generating possibly another "requisition" for the personnel. Finally, a good headhunter would interview the HA to determine if he/she is dissatisfied in their current position and if so, requests they submit a resume to the recruiter for future placement. Therefore, you can see how that one call by the recruiter could be very productive. An ambitious recruiter desiring to succeed would normally plan to make 20 to 30 such calls per day in addition to listening in on a three-way HA–candidate interviews.

Technical Career Survival Handbook. http://dx.doi.org/10.1016/B978-0-12-809372-6.00059-1

What do you need to know about working with a headhunter? Here are some inside secrets:

They make mistakes. All too often the recruiter does not have a thorough knowledge of the candidate's background and/or the requirements of the job. This occurred in my case after being interviewed face-to-face by the recruiter in the Chicago O'Hare airport. He later called me and told me I was not right for the position of manager of development engineering. A few weeks later, he called me up and said his client company wanted me to fly to Denver for an interview with the HA. I was careful not to ask why I was turned down previously; however, after a two interview trips to Denver, I was hired.

Do not count on them. You may get a call and the recruiter explains what sounds like a great opportunity. It may end there if the recruiter cannot convince the HA to conduct a telephone interview or perhaps decides that they cannot pay the fee. Similarly, if you submit your resume to a recruiter, it goes into a data bank and may sit there for months/years before any action occurs.

Timing is everything. When you are laid off, you will be wondering why recruiters do not call you when you need them. Recruiters do not focus on placing candidates who are unemployed. In fact, they would prefer placing an employed person, which lends credibility to the candidate's value. Sometimes, explaining why a candidate is unemployed raises too many questions. If you think your job is in jeopardy of ending in the near future, go ahead and call that recruiter whom you talked to a year ago rather than wait until the cutback is announced.

Be prepared for numerous telephone calls. Recruiters use telephones to collect resume information and pitch a job opportunity thus avoiding bringing candidates in for face-to-face meetings. That saves time and money. Although it is common practice to conduct a face-to-face meeting for the final, they make an offer to meet earlier, particularly when it is advantageous for the candidate to tour the work environment and/or see the local area.

The recruiter is paid by the employer. Obviously, the recruiter has to be certain that the employer is willing to pay a fee to land a candidate. Further, the recruiter has an incentive to encourage the hiring authority to offer the highest annual salary possible since they will likely be paid on a percentage basis. A seasoned recruiter will likely know approximately what the candidate's salary requirement is, but will not reveal that information to the client company. The recruiter has to be careful that the salary requirement is not a deal breaker.

Recruiters depend on referrals. When recruiters call a candidate, they may not indicate how they received the candidate's name and contact information. This creates a slight mystery about the process. Then if the candidate declines to show interest in a proposed position/company, the recruiter will likely ask if the candidate knows someone who might be interested and requests their contact information. Further, the recruiter may ask if the candidate knows of any open positions within the company and to identify the hiring authority.

Headhunters may be essential to your survival particularly after you have gained years of experience. You never know when you may need one, so stay in touch.

Song: "Mr. Tambourine Man"
As popularized by: Bob Dylan
1969

Staffing Firm

60

Unlike a recruiter, a staffing firm, a.k.a. a body shop, provides technical personnel on a temporary, contract, or permanent basis. While employed with Day and Zimmerman, consulting engineers in Richmond, Virginia, we were unsuccessful extending our alliance contract with Honeywell Corporation. However, I had an inkling that our days were numbered even though I had been assigned to Alstom Power and was physically located in their engineering office. Knowing fully well that the Alstom work was winding down, I began to search for my next opportunity.

Staffing firms often align themselves with specific clients, and in some cases will even establish an office within their client's office if they anticipate frequent hiring needs. Since my work with Alstom Power was specific to the Power industry, I contacted a staffing company I found on the internet, which was recruiting technical personnel for an upcoming project for Dominion Virginia Power. The staffing company was experienced in providing technical personnel in the petrochemical, chemical, pharmaceutical, manufacturing, telecommunications, power (nuclear and fossil), pulp and paper, and environmental industries.

During the negotiations, the staffing company made it clear that the position they were contracted to fill was a 10–12-month assignment. Ensuing discussions with them went well leading to an interview with Emil Avram, Dominion project manager (hiring authority, HA) assigned to the power plant to be constructed along the Potomac River. Bottom line, I received a favorable offer, I accepted, and reported for work at the job-site in short order. I described the job-site environment in a previous section of this book. While I was pleased with how the staffing company proceeded with my placement, I had no subsequent contact with them other than receiving my weekly paycheck. My compensation was strictly based on an hourly rate with no benefits.

Here are some of the typical services provided by staffing firms:

- They support client companies' goals and objectives. They work to ensure clients' needs are modeled, analyzed, and implemented accurately.
- Many are industry specific and take care of all the day-to-day administrative and management functions for your temporary workforce, freeing the client company's staff to focus on core business tasks.
- Many focus on hiring specialized technical personnel for seasonal plant shutdowns or outages.
- While working under the direction of the client company, a staffing company employee may be located at the client's site.
- The staffing firm may provide health insurance, 401K, term life insurance subject to negotiation.
- Some claim to vet candidates by conducting thorough interviews, background checks including drug screening, criminal checks, and reference calls. Some test candidates for their expertise using online services.

Technical Career Survival Handbook. http://dx.doi.org/10.1016/B978-0-12-809372-6.00060-8

- They maintain all required employment records and handle the payment of wages, payroll and unemployment taxes, worker's compensation insurance, and proper payroll deductions.

Over the years, you will receive inquiries from recruiters and staffing firms. Maintain a file of those, which specialize in your area of expertise. You never know when you may need to connect with them to hire an employee or find a position for yourself.

Song: "Taking Care of Business"
As popularized by: Bachman-Turner Overdrive
1974

Networking/Newspaper/Internet 61

What do networking, newspaper, and the Internet have in common? They are all a means to finding a technical position. In my case, I found four positions via the newspaper, three positions via networking, and one position through the Internet during the course of my career.

Aside from networking, the newspaper and Internet led me to contact staffing firms that ultimately led to my employment. Networking, on the other hand, led me to contact hiring authorities who were in a position to invite me to an interview to discuss an open position.

Networking: This can be conducted in two ways. First and most common, by using social media, such as LinkedIn, Facebook, and Craigslist. Many of these sites will ultimately lead to staffing firms. Professional networking sites allow you to connect with people you know and the people they know. It is the modern form of business cards. Be cautious when you sign up for online social networking sites, you will be in the public domain.

Secondly, start by targeting a company or industry where you think you might want to work. Company websites will often reveal the names and positions of company personnel. Then, contact the hiring authority directly and explain that you heard that they were highly successful in their industry and you thought they might be in need of someone with your specialty. Some employers even have career pages and invite visitors to fill out candidate profiles, background, interests, and salary requirements.

Newspaper: Companies' direct job postings are rapidly disappearing from the newspaper. In fact, there are more ads in the pet section of my local paper than job want ads. Instead, newspaper job ads are placed by staffing firms that run blind ads for jobs within client companies. So until you contact the staffing firms, you will not be able to discern what company actually has the open position. However, you should not be concerned as long as you have the opportunity to throw your hat in the ring. They will likely request you to send your resume to them often before they reveal their client company's name.

As an alternative to actually reading the physical newspaper, go online and find the newspaper's online job postings. The advantage is you can check with many different locations as long as you can determine the name of the newspaper using an Internet search engine.

Internet: I like the idea of building your own website that essentially posts your resume. While I have not done that with my resume, I had a similar experience selling a historic house. All the information regarding the house—photos, taxes, room sizes, type of heat, history of the house—were included on a website designed just for the house. So when I ran a newspaper ad or talked to someone about purchasing of the

Technical Career Survival Handbook. http://dx.doi.org/10.1016/B978-0-12-809372-6.00061-X

house, I simply provided them with the web address. I sold the house through that medium without involving a broker.

Have you ever Internet searched yourself? You might want to try just to be sure what comes up is favorable in case a potential employer does the same thing. You may have to do some damage control.

A more conventional approach is to visit websites such as Monster.com that post open positions. But do not be surprised if either the job does not exist any longer or has not been retracted. Some industries may have national or regional websites where you can find jobs in your field. A word of caution: it is not uncommon for a company to post a job opening as a fishing expedition to see what talent and salary requirements currently exist. They might have an internal candidate whom they are considering for a position, but want to make comparisons with outside job seekers. In that way they can evaluate the cost to hire versus promotion from within. Additionally, many companies have a "career opportunities" icon you can click on and either apply or see what positions are available.

Song: "Good Riddance"
As popularized by: Green Day
1997

Offers/Counteroffers

Generally a company will make an *offer* only when they are certain that (1) you are qualified and highly likely to accept the position and (2) the compensation package is appropriate. The interview process is where the potential employer gains this comfort level.

When I was interviewed for the position of manager of development engineering with Sundstrand Fluid Handling, it was a three-step process. My first interview was held with the headhunter at an airport. Unfortunately, he misread my suitability for the position and turned me down. The second step occurred when the client company reviewed my resume and requested I fly out for an interview. The third step required me to return for an interview except that this time my wife was requested to accompany me. Why, simply because they knew that my decision would include my bride's input on whether she was prepared to move across country and land a suitable position in her field. My second trip was handled very wisely by my future boss. His wife had prearranged casual interviews for my wife with the local department of education and their handicapped children's special school personnel. She also showed her several housing developments in the area. The conclusion, my wife came away with confidence that should I receive an offer, the job and housing possibilities for her were quite favorable. I got the offer.

Ironically, 10 years later, I had the option of being laid off via downsizing or transferring to another division of the corporation on the West coast. She refused to go. So we moved back to the East coast where I started my short alternative career as a headhunter.

Often an offer of employment is made first on a casual basis in a discussion, then later followed by a formal written offer. I recommend not fully committing during the verbal offer discussion but, of course, you should convey your interest and encouragement to the hiring authority. By delaying your response you will be able to:

- Look at all the advantages and disadvantages.
- Discuss with family.
- Compare the offer to your current situation.
- Consider relocation expenses if any.
- Discuss with your current boss.

Yes, in some instances you may want to use the recent offer as leverage for a raise or promotion with your current employer. This must be approached very delicately; it is a gamble but can be effective as a wakeup call to your boss who may have been taking you for granted. But be prepared for the worst—getting fired on the spot; it happens.

How long will it take to get an offer? This is difficult to predict but if you can determine the reason for the opening, it may give you a clue as to the timing. If a company

Technical Career Survival Handbook. http://dx.doi.org/10.1016/B978-0-12-809372-6.00062-1

is replacing a quit for a critical position, they may react quickly. If it is a newly created position, the hiring process may drag on. But a further indicator may be the health of the economy, according to a study by Dice-DFH and their *Vacancy Duration Measure* index. Since the great recession, employers are taking longer; on the average 25 days to fill a vacant position. That is a 13-year high. And for large companies, the duration may be double the working days to fill. According to a survey by CareerBuilder, a leading job board, managers' fear of making the wrong decision have slowed down the hiring cycle significantly. This can be very frustrating when you are a job seeker.

Now moving on from the offer, what is a *counteroffer*? It is the headhunter's worst nightmare. Let us assume you have gone through the interview process and the future company makes an offer of employment. You have the offer in hand. Hence, the next step is to face your boss and tell him you are resigning effective, say, in the customary 2 weeks. Then surprisingly, he indicates that you are a real asset to the organization and they have a promotion that they were going to present to you within the next week or so along with the associated salary increase. What do you do next? You can no longer compare the offer of employment with your current position, instead, you will need to evaluate the promotion position.

Now when you inform your recruiter about the counteroffer, he will undoubtedly discourage you from accepting it. His argument will be something like: "if they were really serious, they would have already promoted you." Further, he will try to convince you that you are now known to be dissatisfied with the company, you are "toxic," and there is no guarantee they will make good on their promises. This is certain to be a difficult decision for your survival, and in a later section I discuss a decision-making procedure that is applicable to this dilemma.

Song: "Five O'Clock World"
As popularized by: The Vogues
1966

Part IV

On the Job

The Midcareer Years

63

During the first phase of your career, you did not know what you did not know. Since that time, you have matured in your profession, possibly had a job change or two, built a solid resume describing your discipline, industry, and specialty, and now are in a different phase of your career, which I will refer to as "the midcareer years." In this phase, you know what you do not know and you will hone your skills and develop expertise in your field. This will not necessarily occur academically, but through focusing on other aspects of your career, such as joining a trade association or professional society, preparing technical articles, presenting papers, attending seminars, obtaining a professional engineers license, preparing budgets, goals, and objectives, and scheduling projects. This may or may not result in a supervisory or management role, but perhaps a senior or principal engineer position with your company.

What are some of the characteristics of technical personnel at this phase of their career? In Chapter 27, I discussed the technical spectrum identifying a range of dependency on academics in the performance of job specialties. The spectrum included a range of specialties from research to sales. Regardless of their specialty, there are three common elements of success for technical personnel: creativity, productivity, and ingenuity. Let us look at them separately.

Creativity: One might associate this characteristic with strictly highly introverted design and development personnel where "newness" of a product, process, or concept would seem to be the emphasis. I contend that creativity is applicable to the entire technical spectrum including sales. It can occur by chance, by plan, or spontaneously. In my previous position as manager of development engineering, we would often use the "dumb idea" or "brainstorming" approach to create novel concepts for products.

An engineer would come up with, say, 10 dumb ideas, show/sketch a physical configuration, and list the pros and cons for each. Also, if there was a rough idea of the cost magnitude, that would be listed also. The objective was to not rule anything out until all ideas were defined. When completed, we would step back and discuss them one by one. Invariably, there would always be one great idea among the batch that would be apparent after using a decision analysis technique. I will describe this in a later section. We would then take that one great "dumb" idea and run with it. I use this only as an example, but the same process is applicable to any of the specialties in the technical spectrum.

Productivity: In a previous section, I described how Sundstrand Fluid Handling (SFH) was reorganized by creating three *separate* and unique product groups. The reason was to increase productivity or "output" of each product line. In that example, it resulted in a very effective change. Obviously, in a different or opposite example, one might find that *combining* three product lines under one organization improves productivity.

In Chapter 24, I described how at SFH we purchased computer software to design aerodynamic, high-speed rotating impellers for a multistage air compressor. To accomplish these designs by hand calculations would be extremely time-consuming and less accurate. Utilizing the software, many different computer runs made it possible to arrive at the "optimum" design.

While employed by General Electric Corp. Gas turbine Division, I saw how computerized schedules improved productivity by breaking down hundreds of tasks required to design, manufacture, assemble, test, and ship an industrial gas turbine based on a one-year delivery time period. This was my first encounter working with computer-generated schedule dates. I was also surprised to learn that a senior level technician was devoted full time to generating schedules.

Ingenuity: Problem solutions that are thought to be clever, resourceful, and inventive are often said to be ingenious. They are often derived from one's experience, adaption, modification, or application of acquired knowledge to the solution of a new problem.

At Hamilton Standard Division (HSD), aircraft auxiliary equipment such as compressors, turbines, and fans had been historically produced in pairs, such as turbine-driven compressors, turbine-driven fans, and even electric motor–driven fans, to provide cabin air conditioning. When these were used in combination for, say, an onboard environmental control system (ECS), the combined weight and cost was a significant problem. For the Boeing 747 aircraft ECS at HSD, we utilized an ingenious design nicknamed the "three wheel machine."

This package utilizes bleed air from the engines or ground cart directed into a primary heat exchanger. Outside ambient air is used as the coolant in this air-to-air heat exchanger. Once the hot bleed air has been cooled, it is compressed through the compressor impeller. This compression heats the air and it is directed to the secondary heat exchanger, which also uses outside air to reduce the temperature. The temperature of the compressed cooled air is somewhat greater than the ambient temperature of the outside air. The compressed air then passes through the turbine wheel that extracts work from the air as it expands, cooling it to below-ambient temperature. The work extracted by the expansion turbine wheel is transmitted by a shaft to rotate both the compressor and fan impellers, which draw in the external air for the heat exchangers during ground running; ram air is used in flight. Finally, the fan impeller draws the cooled air through the aircraft cabin to the individual air vents.

That approach resulted in the combination of turbine, compressor, and fan impellers on one shaft, thus reducing both weight and manufacturing cost of the system per plane. Although designing the package was tedious, the payback was well worth the effort. Every ounce of component weight saved has a significant assigned dollar value in aircraft systems design.

Song: "Bang on the Drum All Day"
As popularized by: Todd Rundgren
1983

Technical Library

64

Essential for the tool kit of technical personnel is to build and utilize a technical information library. A former coworker of mine at General Electric used to refer to it as his "smart book." By that, I am not necessarily referring just to books, but any material that includes information pertaining to one's discipline, industry, or specialty. I do not recall when I began to collect technical library articles, probably when I began to receive trade magazines, such as *Plant engineering*, *Plant Services*, *Turbomachinery*, *Consulting Engineer*, *Chemical Engineering*, and *Machine Design*, that were largely free if you are qualified.

Often, someone in the office would attach an article to an email and forward it to others in the department. If it appeared relevant, I would save a copy. Then I began to include special calculations, reports, photos, competitive product analysis, and procedures to the mix—anything that I thought might be of value to help me survive in the future.

You may want to consider setting up a filing system when you have accumulated so much information that you have trouble keeping track of it in your head. I would recommend that you select several major headings that pertain to the entire realm of subjects you anticipate collecting in your library. Once you have created the list, arrange them in alphabetical order. Assign numbers to them such that all the articles under "C1," say for compressors, are numbered consecutively from 1 to whatever. Then for condensers use "C2." So that all the articles for compressors would be C1.1, then C1.2, and so on, and all the articles for condensers would be C2.1, C2.2, and so on. After you have started the system, put labels on each article in the top so that they are highly visible when thumbing through a file folder. Lastly, create a list by article number and subject and include a brief statement pertaining to the topic of each article so that you can look them up conveniently in the future.

Once you have established a technical library, you will be surprised what a valuable survival tool it will prove to be. You may even find that you will be able to pull up useful information you may want to pass on to your peers or even your management. Years ago, my general manager once asked me in the hallway "what is reverse osmosis"? I wish I had saved an article on the subject so I could have provided him with the answer on the spot, but I had to research the subject and get back to him later. Ironically, few years later we were supplying pumps into the reverse osmosis water purification system market.

Song: "Rocket Man"
As popularized by: Elton John
1972

Technical Career Survival Handbook. http://dx.doi.org/10.1016/B978-0-12-809372-6.00064-5

Associations/Societies

65

Some professional organizations are comprised of many members with the *same discipline* and are referred to as a *society*, such as the American Society of Mechanical Engineers (ASME). The membership may be comprised of individuals from a variety of industries, but the primary discipline would be generally common to all. ASME "promotes the art, science, and practice of multidisciplinary engineering and allied sciences around the globe via continuing education, training and professional development, codes and standards, research, conferences and publications, government relations, and other forms of outreach." The same is largely true for American Society of Electrical Engineers, American Society of Civil Engineers, Society of Automotive Engineers, and many others.

These societies generate standards, conduct research and development, lobby for legislation, provide training and education, and collect dues from members, but are nonprofit organizations. They are national organizations with staff that manage, organize, and oversee member activities.

There are also professional organizations comprised of a membership from *different discipline backgrounds* but with a common purpose they are also societies. Examples of these societies are: American Society of Education, American Society of Women Engineers, American Society of Hispanic Professional Engineers, National Society of Professional Engineers, and Tau Beta Pi Engineering Honor Society.

A common activity of most societies is the preparation and distribution of "standards." A standard is a set of guidelines that serve as instructions for designers, manufacturers, and users of various types of equipment. Often, standards become "Codes" whereby they are adapted by government agencies as enforceable laws. Standards are derived from input by many companies contributing to the overall effort of a designated committee of volunteers from the membership.

Professional organizations comprised of many members of the *same industry* are known as industry associations, for example, the Compressed Air and Gas Institute (CAGI) that works in coordination with other standards organizations, including Pneurop (combination of pneumatic and Europe) and the American National Standards Institute. They were founded in 1915 and represent manufacturers of compressed air system equipment including air compressors, blowers, pneumatic tools, and air and gas drying and filtration equipment. Like societies, they are a national organization with staff that manage, organize, and oversee member activities. Representatives to CAGI are multidiscipline but are engaged in the manufacture of the said products. Unlike societies, industrial associations privately collect and assimilate sales data submitted by member companies. The data is presented in a generic way so as not to identify individual member sales data. It is then distributed among the members so that they can determine their market share of the industry as a whole.

Technical Career Survival Handbook. http://dx.doi.org/10.1016/B978-0-12-809372-6.00065-7

While employed as manager of development engineering, I represented my company in the CAGI activities. We were field-testing a multistage industrial air compressor product line and wanted to have input on a standard that was to be developed for equipment such as ours. We met several times to prepare the standard that became available to the public for a small fee. The standard included terminology, testing procedures, ratings, arrangements, and performance conversion information. This standard did not become the basis of a "Code." Often, technical papers are presented by association members at their annual membership meetings. Usually the information covered is mostly technical in nature without divulging company proprietary information.

As a personal benefit, becoming a representative to an association is an excellent way to network within the particular industry in which you are employed. You will expand your scope of the industry and make valuable contacts.

Caution: Be careful about having discussions that could be considered "price fixing." This occurs when participants collude to sell a product, service, or commodity at a fixed price, or by controlling the supply.

Song: "Cherish"
As popularized by: The Association
1966

Seminars/Webinars

66

In Chapter 36, I explained the importance of training in order to realize growth in your specialty which will ultimately lead to advancement and recognition in your company and industry. Today, there are many opportunities to attend group seminars and online webinars.

Here are two types that are offered:

Leadership: The Dale Carnegie Institute offers a wide range of courses three ways—in office, online, and off site. Courses include leadership development, presentation effectiveness, presenting complex information, public speaking mastery, and many more. They are famous for *How to win friends and influence people* course which I took many years ago. Off-site seminars may cost several thousand dollars, so you may want to consider in-office training and spread the cost for a trainer over several people.

Technical: Many technical courses are available online for low cost that allow you to earn *continuing education units* (CEU). One CEU is equivalent to 10 contact hours of instruction or study in an organized class conducted by a qualified instructor. In the engineering community, a *professional development hour* (PDH) is commonly used. More on that later. Examples of people who need CEUs are teachers, architects, engineers, emergency management professionals, and others. Short courses for engineers include thermodynamics, fluid mechanics, electrical circuits, heating and cooling systems, noise control in buildings, and many others.

Previously I mentioned engineering software, such as ANSYS (heat transfer analysis—mechanical engineering), STRUDL (structural analysis—civil/structural engineering), PSpice (circuit analysis—electrical engineering), and CHEMCAD (process analysis—chemical engineering). These programs are complex and require several hours or even days of training to become proficient. Therefore training is generally conducted at the software developer's training facility where each person will have access to their own computer with the latest version of the software downloaded. Many PDHs are awarded upon completion of these seminars.

While working for Day and Zimmerman Engineering, we would conduct short courses provided by technical personnel within the company, often during lunch. Because it is important that engineers have a broad understanding of various disciplines within their own organization, we would have for example, a controls engineer present methods for maintaining a liquid level in an evaporator. The presentation would be made to chemical, mechanical, and civil engineers, and designers. The presenters would keep it short and simple and use illustrative examples. We called it "cross training," yet another tool for survival.

Song: "Last Thing on My Mind"
As popularized by: Tom Paxton
1978

Technical Career Survival Handbook. http://dx.doi.org/10.1016/B978-0-12-809372-6.00066-9

Publishing

67

There is an expression "publish or perish" that is often used when referring to professors, particularly at the graduate-school level. While engineers, scientists, and technicians in industry are generally not evaluated based on their publishing efforts, there are advantages that may not be obvious.

Often articles appearing in trade or industry magazines or journals are authored by subscribers of the publication, consulting engineers, and equipment operators/users. These articles address technical solutions, products, product features, or calculations that are relevant to the magazine subject matter—equipment, chemical processes, instrumentation, or structures. Examples of these trade magazines are: *Aviation Week & Space Technology, Biotechnology and Bioengineering, Electronic Design, Energy & Environment, Journal of Fluid Mechanics, Machine Design, Chemical Engineering*, and many others.

Commonly known in academia as a thesis, *papers* are prepared, published, distributed, and sometimes presented and defended in a technical conference or forum. In industry, papers are prepared and presented at conferences conducted by societies or associations. The technical information contained in the papers may eventually lead to the development of industry standards. There are four reasons why it is important for technical personnel to prepare magazine articles or papers:

1. *The process of preparing articles or papers is a unique learning experience for the author(s).* In order to prepare technical articles, it is necessary to conduct a certain amount of research, calculations, testing, and organizing them in a coherent readable format. This may be accomplished by one individual or even a team of technical personnel. A test facility may be necessary to prove a phenomena or performance prediction. A judgment must be made as to whether revealing the subject matter makes good business sense: Does it divulge company proprietary information? Does it give the competition an unfair advantage?
2. *They provide company product recognition.* During the course of developing a multistage packaged air compressor at SFH, we utilized a finned tube, water to air inter stage cooler manufactured by a sister division of the corporation. The design was the same as used in the mass production of air conditioning equipment and was therefore relatively inexpensive. We needed a calculation procedure to convert the performance of the coolers from one set of cooling water and air conditions (in the lab) to a different set of conditions (the customer's site).

I identified the necessary equations to make the performance conversion using dimensionless numbers and we incorporated the procedure into a computer performance prediction program for the air compressor. This program enabled us to quote the air compressor performance based on the customer's site conditions: temperature, atmospheric pressure, and humidity. I decided to prepare a technical article describing the prediction procedure which was later published in Chemical Engineering

Technical Career Survival Handbook. http://dx.doi.org/10.1016/B978-0-12-809372-6.00067-0

magazine. We purchased reprints of the article and passed them out as sales literature in order to promote the product line. The article served to provide credibility to the product as we began to penetrate a market for our new line of air compressors. Unfortunately, after several years of production, a management decision canceled the product line due to the high cost to manufacture the product in spite of several attempts at cost reduction.

3. *They provide individual recognition.* While employed by SFH, we developed a high speed pump for the American Petroleum Industry (API) that operated in a high-flow region that previously had not been available. We achieved this performance utilizing an axial impeller seal that operated with only a few thousandths of an inch clearance while rotating at speeds up to 8000 rpm. The seal was later patented. I compiled a technical paper pertaining to the design and performance of the new product being careful not to reveal company propriety information. I submitted the paper at the annual Industrial Energy Technology Conference and exhibition in Houston, Texas and I was asked to present the paper at the meeting. I gave thanks and credit to our marketing services group. They did an excellent job formatting the paper to include many illustrations and cutaways making it a very successful and memorable effort.

4. *The articles provide free advertising.* Trade magazines like those mentioned in the preceding paragraphs typically solicit articles from companies offering related products. I prepared technical articles for several of my employers that appeared in publications and reached our target market readership. Not only were the articles published at no charge, often the publisher would a pay an honorarium to the contributor. Concurrent with the article, usually two to three pages describing the subject product, my employer would then place a subtle ¼ page paid advertisement adjacent to the article so that the reader knew where to purchase the product or service featured in the article. We were successful in generating many sales leads in this manner.

Caution: be certain to obtain company approval for any written or oral material presented to be certain trade secret information is not conveyed. Additionally, beware of revealing any confidential marketing data and/or strategies to the public. Also be prepared to "defend" your publications as you will get questioned on the content particularly if it covers new or controversial material.

Song: "Glory Days"
As popularized by: Bruce Springsteen
1984

Industry Standards

68

One of the best tools technical personnel have available are specifications, codes, and technical standards or, as they are sometimes called, "advisories" or "recommendations." I will refer to them collectively as standards and there are hundreds in use today. They serve as an established norm or requirement in regard to technical systems, equipment, and/or materials. Standards may be developed privately or unilaterally, by a corporation, regulatory body, military, and so on. They can also be developed by groups such as trade unions or standards organizations. They are prepared in draft form initially, reviewed by an oversight committee, revised, and then published as a formal document. They must be revised and updated periodically. Standards present established uniform engineering or technical criteria, methods, calculation procedure, test, calibration, processes, or practices.

Standards serve as:

- An aid to procurement of standardized equipment and materials.
- A primary standard usually under the jurisdiction of a national standards body.
- A reference in a metrology system for traceability and paper trail of calibrations back to a standard or reference.
- Mandatory regulations if adopted by a government or incorporated in a business contract.

A "standards organization" I am most familiar with is *The American Society of Mechanical Engineers* (ASME). The organization wrote the famous Boiler and Pressure Code in 1911 in order to protect the public. After the conception of the steam engine in the late 18th century, there were thousands of deaths attributed to boiler explosions. So a group of volunteers formed a committee and used their expertise to develop the code with rules for design, fabrication, and inspection of boilers, piping, and other pressure vessels.

Another ASME standard that I am familiar with is titled Safety Standard for Mechanical Power Transmission Apparatus, originally drafted back in 1927. The stated the purpose of the standard "is to provide guidance for minimizing the likelihood that people will incur injury when in the proximity of mechanical power transmission apparatus." This standard is written in a "performance" mode rather than a "specification" mode to encourage the appropriate use of ingenuity and imagination in achieving a maximum degree of safeguard. Interestingly, the committee that created the standard consisted of members from insurance companies, Department of the Navy, consultants, the American Society of Safety Engineers, and equipment manufacturers.

Does the standard eliminate accidents with power transmission equipment? The answer is no, but it "suggests" ways of avoiding them in most cases. For example, the standard spells out machinery guard dimensions, height of barriers, safety marking means, warning signs, and motion hazard devices.

Technical Career Survival Handbook. http://dx.doi.org/10.1016/B978-0-12-809372-6.00068-2

A standard that my friends in the electrical department utilize extensively is the National Fire Protection Association (NFPA) National Electric Code or NFPA 70. This Code has been adopted in all 50 states and serves as a benchmark for safe design, maintenance, inspection, and operation of electrical machinery. It is intended to protect personnel and property from electrical shock and as a potential ignition source of fires and explosions. It also provides guidance on how to minimize the propagation of fire and explosions due to electrical installation hazards associated with electrical energy. The standard consists of three main sections as follows.

1. *Safety related work practices*: general requirements and safe work practices in an electrically hazardous area.
2. *Safety related maintenance requirements*: when working with substations, switch gear, motor control centers, wiring, fuses, circuit breakers, batteries, portable electric tools, and equipment. Use of the proper personal protective equipment is addressed.
3. *Safety requirements for special equipment*: work practices for use of electrolytic cells, batteries, and battery rooms and lasers.

This standard does not include, however, safety related work practices for ships, aircraft, underground mines, railways, communications equipment, and equipment owned or leased by an electric utility company.

Finally on this subject of major importance, the Occupational Safety And Health Act (OSHA) became law in 1970. Under this statute, rules are presented to provide a safe and healthful workplace by setting and enforcing standards, education, training, and assistance. Employers must comply with the Act and keep their workplace free of serious recognized hazards. I recall that when this Act was passed, it was rumored that considerable emphasis would now be placed on obtaining professional engineer's licenses as a means to review, approve, and "stamp" design documents, particularly those that pertained to equipment, structures, and facilities that were to be used by the public. More on that subject in Chapter 72.

Song: "Safe and Sound"
As popularized by: Capital Cities
2011

Employment File

This subject practically falls in the category of housekeeping, but an employment file is a basic tool that should not be overlooked. From the time you first begin your employment all the way to retirement, it is recommended that you organize a system of files to collect information essential to your employment history. You never know when you will need to refer to a past projects, coworkers, or achievements. Here are some categories that you may want to consider in your files:

- *Companies*: collect brochures or articles describing your employer's products and/or services. Organization charts should be kept as it may be difficult to recall who reported to who.
- *Reports*: stash those reports that you deem particularly representative of your work or your team's work in the file.
- *Calculations*: if there are any that were particularly complex or that required derivation? Many companies utilize a standard format which includes review and approval sign offs.
- *Communications*: letters, emails, or fax messages that provide a significant direction or response to an inquiry that you might have made. Maintain copies of your past performance reviews.
- *Photos*: they are worth a thousand words and often serve to document an event or result of a project or product development.
- *Business cards*: yes, they are still handed out by most personnel that you will meet. I like to write the date which I received it on the card.
- *Articles*: include papers and magazine articles that you have prepared and/or presented.

Obviously the above information can be extremely useful when you participate in a job interview, when being considered for a promotion, or for your survival. Call it a portfolio if you will. The file will also help when you are trying to recall what has been done in the past so as to avoid reinventing the wheel.

Song: "Mrs. Robinson"
As popularized by: Simon and Garfunkel
1968

Technical Career Survival Handbook. http://dx.doi.org/10.1016/B978-0-12-809372-6.00069-4

Patents

One of my responsibilities at SFH was to manage all the patent activities. We were primarily concerned with "utility" patents that were new, nonobvious inventions pertaining to our product line of pumps, compressors, and blowers. However, many companies pursue "design patents" that are new or original designs pertaining to appearance only. My responsibility included recognizing opportunities for applying for patents, authorizing patent searches, reviewing "prior art," and determining the value in pursuing a patent application. In Chapter 67, I described an axial impeller seal configuration we developed at SFH for a high-speed, high-flow API market. After we had successfully tested the seal design in the laboratory and the field, we took the following steps to patent the invention.

First, we had to be certain that we were not infringing on an existing patent, so we authorized our patent attorney to conduct a patent "search." The search requires a review of existing patents contained in the United States Patent and Trademark Office (USPTO) in order to determine if we were infringing on any previously patented rotary seal designs. This process required several weeks and resulted in not finding any similar patented seal designs.

Second, we reviewed competitive sales literature to determine if there was "prior art" that was not necessarily a patent-protected design but existed in the marketplace regardless. The seal configuration was also determined to be nonobvious, another important criteria. There is no sense in applying for a patent if the invention is already in the public domain as prior art. The wheel comes mind.

Third, based on no possibility of infringement, no prior art determined, and a high likelihood we would market pumps featuring the invention/seal, we authorized our patent attorney to draft a patent disclosure and make application—a process that took several months.

There are five main parts to the patent application:

1. Petition, the request.
2. Specification, the description.
3. Claims, the protections.
4. Drawings, the technical details.
5. Oaths, verification that you are the first inventor.

Following a description of the invention or specification referred to as the "abstract," the most critical portion of the patent are the "claims." In this section sentences are constructed to very specifically describe how the invention is different than any previous, similar devices. The number of claims may range from 1 to 20, which corresponds to narrow to broad protection against competitive inventions. Most patents have less than 10 claims. Claims are very carefully worded to describe the elements of the invention

Technical Career Survival Handbook. http://dx.doi.org/10.1016/B978-0-12-809372-6.00070-0

that are to be protected by the patent. So it is important that the patent attorney has a good working knowledge of the invention which may necessitate having considerable discussions with the inventor(s). The USPTO requires that claims be:

- Very specific
- Clear
- Stated in one sentence
- Distinct from other claims in the patent application
- Consistent with the patent application narrative

Rarely does the USPTO issue "letters of patent" upon the first application. They will usually issue an "action" which will include the patent examiner's objections to the claims. In response the patent attorney must issue an "amendment" in which he further explains, modifies, or drops the claims. This results in narrowing the scope of the patent. The examiner may totally reject the claims, and you may appeal. If the appeal is rejected, you lose.

The USPTO designates five guidelines/questions as a basis for granting a patent as follows:

1. Is the invention original and practical?
2. Is it unknown?
3. Is it useful?
4. Is it workable?
5. Is it past the concept stage?

The pump seal patent that we pursued was classified by the USPTO as *general and mechanical*, the most common. Other categories include *chemical*, and *electrical*. The term for these patents is 17 years. You may hear the term "design patent" used occasionally. This refers to a patent based strictly on appearance, such as a table, chair, glass, handle for a shovel, and a tire tread. The term for these patents is 14 years.

Before even thinking about applying for a patent, it is important to decide how you plan to market the product. Is the patent necessary to enter the marketplace and be competitive? The patent process is time-consuming and expensive therefore be certain patent protection is required.

An alternative to utilizing a patent on a production basis is licensing it to others in return for royalty payment(s). A famous example is the inventor Robert Kearns and his "blinking" (intermittent) windshield wipers patented in 1964. Since Kearns, a prolific inventor, did not own an automotive manufacturing company he tried unsuccessfully to market the electronic control system to the big three automotive companies. But, surprisingly, within a few years, intermittent wipers were being offered on Ford, General Motors, and Chrysler automobiles. Consequently litigation ensued for many years and the results of the infringement case initially against Ford resulted in royalty awards to Kearns for each wiper system produced, making him millions. Sadly the 12-year legal struggle took its emotional toll on Kearns and finally the patents on the wipers expired in 1988.

Big companies dominate when it comes to patents issued. In my local area of Richmond, Virginia, big companies like Altria (tobacco and food) received 76 patents,

DuPont Company (materials) 35, MeadWestvaco (materials) 26, and Honeywell International (chemicals) received eight patents over the past year. Universities like nearby Virginia Commonwealth University in 2013 received 13 patents, one for a medical inhaler that delivers medications into the lungs more effectively. In fact, our local paper the *Richmond Times Dispatch* publishes brief summaries of issued patents on a weekly basis which makes for an interesting read for technical personnel.

Lesson learned, be constantly aware of potential patentable machines, features, systems, or processes. Often companies will assign employees to a patent committee to determine whether to pursue patent or trademark protection.

Trade Secrets

In many instances, trade secrets, another form of "intellectual property" similar to patents, may be more practical than pursuing a patent application. According to the United States Patent and Trademark Office (USPTO), a trade secret is information such as formulas, patterns, devices, methods, techniques, or processes used in a business that provide an economic advantage over competitors who do not know or use it.

Unlike a patent, a trade secret is not registered with the USPTO, has no expiration like a patent, and protection continues until discovery or loss. Let us go back to the SFH centrifugal pump with the patented axial impeller seal example from the previous section. In order to manufacture the pump/product, dozens of drawings were prepared by engineers and designers describing many unique parts and many that are catalog items such as bolts, gaskets, and O-ring seals. Unique parts are often machined from forgings, castings, or bar stock. Tolerances are assigned to parts providing a nominal dimension with an allowable variation of plus or minus some dimension or "tolerance." Tolerances are determined carefully by doing a "stack-up study" and evaluating the tolerances in the worst case. In some cases, experiments can be conducted in the laboratory to evaluate tolerance effects.

In the situation with the patented axial seal, tolerances were critical. If they allowed too much clearance, the seal leakage might be excessive and/or the rotating member of the seal might hit the stationary member and cause a catastrophic rub. If the tolerances are too tight, the manufacturing cost might be excessive. To protect the tolerance information included on the machining drawings, a note is included on the drawing titled "proprietary design." The note explains that the drawing is not to be shared with those not authorized to review the drawing, namely competitors or vendors that might be inclined to show the drawing to a third party. This is referred to as "misappropriation." In addition to tolerances, calculations and manufacturing techniques may also be considered trade secrets. However, that is not to say that others acting independently might also discover the same tolerances, equations, and techniques as we did at SFH.

Therefore, you see that trade secrets are an alternative to patents based on maintaining closely guarded information. There are, however, risks that an employee will leave the company and take the secret information to a competitor. Also, there is nothing to prevent a competitor from obtaining a "trade secret" product, disassembling, measuring components, and then using the information to create manufacturing drawings. This process is referred to as "reverse engineering."

The decision on how to protect an invention is a business decision (cost vs. benefit) and depends on weighing the relative benefits of each type of intellectual property. Some will say the important consideration is to enter the marketplace quickly and

Technical Career Survival Handbook. http://dx.doi.org/10.1016/B978-0-12-809372-6.00071-2

establish the product as the economical solution to a problem in the marketplace. Additionally, a patent application can be pursued simultaneously while a trade secret is maintained for the first 18 months that the USPTO patent application is underway.

Song: "Do You Want to Know a Secret"
As popularized by: The Beatles
1963

Professional Engineer

In a previous section, I mentioned that with the formation of Occupational Safety and Health Administration (OSHA), it was speculated that emphasis would be placed on professional engineers (PE) stamping (approving) design documents. What does stamping a design document accomplish? It simply represents that the document has been reviewed and approved by someone "competent" in the subject. This is a frequently used term in legal documents meaning that a PE is sufficiently knowledgeable, capable, and fit to render a design that could affect the health, safety, and welfare of the public, satisfactory. Licensing requirements in most states extend to landscape architects, engineers, architects, and land surveyors.

In order to ensure competency, there are many requirements that must be met in order to obtain a PE license. Although requirements may vary among states, here is a generalized list:

1. The candidate must be of "good moral character" and not convicted of a felony or misdemeanor.
2. A fully documented application shall be submitted by applicants seeking licensure, certification, or registration to be received in the licensing board's office no later than 130 days prior to the scheduled license *examination*.
3. Submit written references verifying the candidate's experience. The individual(s) providing the reference must have known the applicant for at least one year.
4. Applicants shall be able to speak and write in English to the satisfaction of the licensing board.
5. Applicants must have completed an accredited undergraduate engineering or engineering technology curriculum of four years or more, or a graduate engineering curriculum, approved by the licensing board.
6. The applicant must have passed the Fundamentals of Engineering (FE) examination and thereby is an "Engineer-in-training (EIT)" and has completed any one of the several combinations of education, or education and experience. Upon passing the exam, the candidate will receive the EIT designation as established by the National Council of Examiners for Engineering and Surveying (NCEES). The FE exam may be taken any time after graduation.
7. The applicant must have "qualifying engineering experience," usually at least four years. Also, the applicant must have a record of *progressive* experience in engineering work during which the applicant has made a practical utilization of acquired knowledge and has demonstrated progressive improvement.
8. The applicant must pass the professional engineer (PE) examination. An applicant will receive his or her license to practice engineering upon achieving a passing examination score as established by NCEES.
9. Pay the applicable nonrefundable fees. Often employers will allow you the time off the job for exams and reimburse you for incurred expenses and fees.
10. To maintain the PE license, the holder must satisfactorily complete 16 professional development hours (PDH) biannually and pay a renewal fee. According to the ASME, one PDH

Technical Career Survival Handbook. http://dx.doi.org/10.1016/B978-0-12-809372-6.00072-4

represents one contact hour of instruction, presentation, or study. A maximum of 8 PDH are earned by participating in an 8:00 a.m. to 5:00 p.m. seminar with a 1 h lunch break.

It is advisable to follow through on the above steps early in your career while past academic studies are reasonably fresh in your mind. But for all that effort, you are probably wondering why deal with it?

First, even if your career does not lead you to PE stamp engineered documents, a PE license serves as a significant credential.

Secondly, it serves as an indicator that you are eager to excel and commit to maintaining high standards indefinitely.

Thirdly, by maintaining your license, you may be considered for future promotions whereby a PE stamp will be essential.

You might make a comparison of a PE license to a lawyer passing the bar exam. By not passing the bar, the lawyer's duties are restricted. In some instances, without it, you may be at a disadvantage by not being able to stamp a document for a client.

Song: "Signs"
As popularized by: Five Man Electrical Band
1970

Safety

In a previous section, I explained how the subject of safety was emphasized on a job site where I was employed. Fortunately, I have not witnessed any serious accidents; only minor cuts and bruises. To avoid accidents in the workplace, personal protection equipment (PPE) is utilized depending on the hazards likely to be encountered. Most companies require classroom training on the hazards to be expected and the appropriate PPE to be utilized. Training is typically based on compliance with the OSHA standards contained in the Occupational Safety and Health Act of 1970. Some companies send weekly safety topics and tips to employees to read and acknowledge a response accordingly.

Testing may be required to ensure personnel are adequately trained. Upon completion of the training, a badge or hard-hat sticker is usually issued to be displayed while occupying the workplace. Company safety personnel often issue certain PPE before allowing entry in to the work area. However, workers must be provided and wear clothing that is acceptable for the expected hazards.

There are five workplace categories where I have experienced major emphasis on safety issues, each with a particular set of hazards. Table 73.1 is a guide as to expected hazards and the appropriate PPE to be utilized.

Most companies having the stated workplaces, usually have a safety engineer, coordinator, or point man designated to ensure safe conditions based on:

- Planning for the prevention of accidents
- Safety training programs
- Maintaining safety records
- Providing PPE
- Reporting accidents and near misses
- Preparing for emergencies

In addition to addressing hazards in the workplace, there are procedures, systems, and equipment that also need to be reviewed for safety considerations. To name a few, electrical lockout/tagout, barricades, ladders, scaffolds, cranes, mobile equipment, work permits, rigging, fire protection, and material handling equipment. PPE and training are important but so is common sense.

Often engineering and construction companies will conduct job hazard analyses (JHAs) to be certain that employees are aware of the hazards of scheduled tasks and the corresponding behaviors to avoid accidents. This is usually conducted at the outset of the project.

Most importantly, when moving about in any plant, always be aware of your surroundings and hazardous objects. Many plants are vast and therefore it is critical that other personnel are aware of your location, particularly if you are climbing scaffolds or towers in remote parts of the plant. Two-way radios can be a useful tool in these

Technical Career Survival Handbook. http://dx.doi.org/10.1016/B978-0-12-809372-6.00073-6

Table 73.1 **Recommended PPE**

Workplace	Hazards	PPE Recommended
Construction sites	Scaffolds and ladders	Harness and hard hat
	Power cords	GFCI
	Noise	Hearing protection
	Rough metal objects	Gloves and steel toe boots
	Airborne dust and dirt	Safety glasses
	Welding and burning	Safety glasses and welding shields
		Nonsynthetic clothing
Chemical plants	Chemical sprays	Eyewash and shower stations
		Goggles and long sleeves
	Hot surfaces	Specialized boots and gloves
	Fumes	Chemical-resistant gloves Respirator
	Pipe racks	Hard hat
	Platforms	Harness
	Noise	Hearing protection
Power plants	Hot surfaces	Gloves
	Steam	Safety glasses
	Noise	Hearing protection
	Pipe racks	Hard hat
	Platforms	Harness
Manufacturing	Machinery	Safety glasses
plants	Rough metal objects	Gloves
		Steel toe boots
	Noise	Hearing protection
	Welding	Welding shields
Laboratories	Chemical sprays	Eyewash and shower stations
		Gowns
	Fumes	Goggles and long sleeves
		Chemical-resistant gloves

situations. I once performed an inspection in a chemical plant of newly installed equipment and instrumentation requiring climbing multiple scaffolds. It did not take long to realize that had I been injured, it might be a long time before help arrived. Take plant safety seriously, it is important for survival!

Song: "I Shot the Sheriff"
As popularized by: Bob Marley
1973

Business Plan

74

Obtain an engineering degree, get a great job, and make a lot of money is not a business plan, they are three vague goals. In order to be a goal, it must be specific and measurable. In order to be a business plan, it must have a specific objective. You may or may not need to prepare a business plan during your career but regardless, you should understand the concept. So let us look at this simple example and restructure it into a business plan.

> **Objective**: Become an established chemical engineer by 2022.
> Goal #1: Graduate with a chemical engineering degree from Virginia Tech by June 2018.
> Goal #2: Get hired by a chemical processing company in Virginia by September 2018.
> Goal #3: Achieve a salary level of $80K by January 2022.

Is your business plan complete now? No, because you have not defined the actions necessary to achieve each goal. Looking at goal #1; what action steps will be necessary? Consider the following action steps:

- Prepare a budget for college.
- Apply and select the college with a chemical engineering degree program.
- Complete 4-year course work with B average or better.

In previous sections, I mentioned the joint venture that brought the Nikkiso manufactured canned motor pump (seal-less pump) line to the SFH product family. Also in a previous section, I pointed out that the General Manager reorganized the company into three product groups so that the introduction of the new product line would receive sufficient emphasis. These were the goals that were in response to a business plan objective as follows.

> **Objective**: Successfully introduce canned motor pumps into the US chemical-processing market by third quarter 1987.
> To achieve this objective, the following simplified goals were developed:
> Goal #1: Sign a joint venture agreement by September 1986.
> Goal #2: Reorganize the company by January 1986.
> Goal #3: Establish a capital budget for the CMP product line by February 1986.
> Goal #4: Complete staff training by February 1987.
> Goal #5: Develop sales literature by March 1987.
> Goal #6: Establish 12 customer CMP field trial test sites by May 1987.
> Goal #7: Build, test, and ship 12 field trial canned motor pumps by July 1987.
> Goal #8: Receive favorable customer testimonials for 12 CMPs by September 1987.

After the business plan was established and approved, individual departments such as engineering, manufacturing, sales, marketing, field service, training, and IT developed their action steps according to the above goals. Goals such as #5 may have only

Technical Career Survival Handbook. http://dx.doi.org/10.1016/B978-0-12-809372-6.00074-8

involved the marketing department. However, goal #6, involved action steps by all departments. Departments were later evaluated based on their performance toward these goals and obviously, individuals assigned to accomplish the goals were evaluated on their participation and achievement of the goals.

Then following completion of the above objective, a new objective for 1988 was developed based on achieving a 30% market share by year end. And again the goals were established to support the new objective. More about budgets associated with goals and objectives later.

Song: "Down on the Corner"
As popularized by: Creedence Clearwater Revival
1970

Schedules

75

Technical projects may be so basic as to involve only one or two individuals for few weeks incurring no capital expenditures. On the other hand, many projects are often extremely complex, require, large manpower, expense, and capital budgets extended over several months in order to successfully complete a business plan. Project managers need tools to manage the project and maintain project schedules and budget commitments.

One of the main tools utilized by project managers is scheduling software. In the previous section, I described some of the goals associated with a single company objective for which multiple departments and a number of technical personnel may be actively involved. Further, a company may have multiple business objectives under way simultaneously. This necessitates the need to breakdown the goals to individual, manageable action steps.

Computer-based scheduling software is available from many suppliers who may also provide user training. Most of the companies where I have worked had a designated scheduler on a full time basis. I indicated in a previous section that General Electric Corp. Gas turbine Division used computer schedules to improve productivity by breaking down hundreds of tasks required to build a single industrial gas turbine in 52 weeks.

Here are a few of the capabilities of most commercial scheduling software:

1. Provides a means of estimating the completion time for a project based on the sum of its subtasks.
2. Breaks down the project into individual tasks each with its required duration, start, and end dates.
3. Provides a "critical path," that is, the longest necessary path (time) through a network of essential activities based on their interdependencies.
4. Defines which departments are to complete specific action steps.
5. Allows compression of action steps thereby identifying overtime or additional manpower if required.
6. Allows management of large programs with multiple suppliers and subcontractors.
7. Identifies specific milestones which may represent payment dates.
8. Reveals the results on the overall schedule of changes in the project.
9. Provides capabilities to assess all aspects of the programs and projects under consideration.
10. Identifies major risk factors such as the impact of cost errors and the effect of contingencies on costs and timing.
11. Provides the insight required to understand the impact of a particularly aggressive schedule.

Fig. 75.1 illustrates an 8 month construction project to complete a building expansion in a chemical plant from start to finish. Although this is a preliminary schedule, it allows estimators to evaluate manpower costs, when various trades are required, how

Technical Career Survival Handbook. http://dx.doi.org/10.1016/B978-0-12-809372-6.00075-X

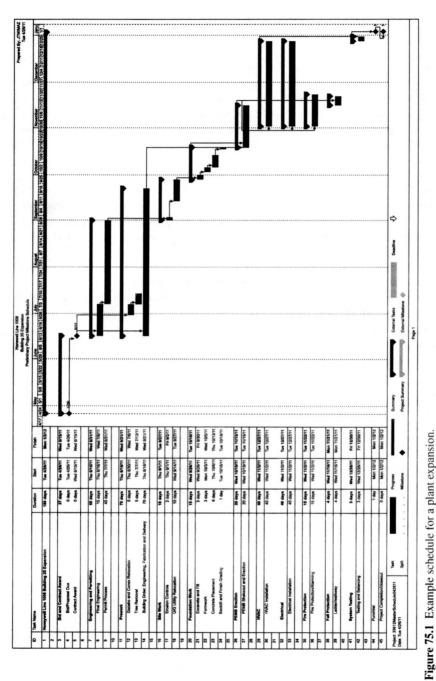

Figure 75.1 Example schedule for a plant expansion.
Source: Honeywell International Inc.

long they will be on site, and when various permits may be required. The schedule also shows activities that are not along the critical path. That is, they can be scheduled for completion without affecting the overall project completion date. For example, see "tree removal" is shown as a 5 day, noncritical path activity.

Note, while this figure illustrates a *project* schedule, scheduling software may be used in a similar way to develop a *product* manufacturing schedule.

Establishing and maintaining schedules on a formal or informal basis can be essential for your survival. But there are two unique problems that occur with any scheduled project that you should be aware of from the outset. Simply put: startup and completion. This phenomenon is depicted on the classic S-curve shown in Fig. 75.2 where the per cent completion is a function of time expended on the project. More problems occur at startup and approaching completion. Whereas once startup problems are resolved, you should accelerate the effort. Then, many would say, Murphy's law will always result in roadblocks as you near the end of the project.

Startup problems result in delays in ramping up quickly. They may include:

Lack of familiarity with the personnel involved and the facility.
Lack of familiarity with similar products or projects.
Delays in obtaining support personnel or software.
Delayed kickoff meeting with the client/customer or project team.
Poorly prepared or vague scope of work.

Completion problems prevent closing out the project on schedule. They may include:

Testing difficulties.
Material delays.
Obtaining client approval of documents/drawings.
Manpower shifted to an alternate product/project.
Last minute configuration changes.

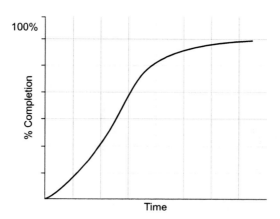

Figure 75.2 S-Curve.

The key to surviving these problems is anticipation and communication. Refer to a later section on meetings for the purpose of reviewing progress.

Song: "Heart of the Matter"
As popularized by: Don Henley
1989

Budgets

76

Chapter Outline

If you think that technical personnel need not be concerned about budgets, sorry, wrong.

Even in the early stages of their career, technical personnel will likely be responsible for completing a simple task or project within a specific time period, a.k.a. man-hour budget.

If you are a supervisor or manager you will likely be required to prepare a budget, function under the constraints of an operating budget for your department, and importantly, your survival may depend on it. Budgets may be based on hours, expenses, or capital investments. Let us look at these closer.

Man-hours

Tasks assigned to technical personnel are generally determined to be completed within an assigned time period often established based on past experience or estimated from known standards of performance. The hours can be readily tracked using time sheets completed by the personnel involved. Additionally, the hours utilized can conveniently serve as a measure of progress toward completion of the project. In other words, if 50% of the allotted hours for the project have been spent, one might say the project is 50% completed.

In reality, man-hours represent an employment expense depending on the individual's salary and benefits known as compensation. However, since the compensation varies among personnel, it is more convenient to refer to man-hours based on an average cost depending on the department. In many engineering organizations, man-hours may be lumped in categories such as engineering, design, factory labor, and/or technician. When the man-hours are viewed in total, they may be referred to as direct labor cost.

Expenses

This budget item is known to all technical as well as business personnel. There are many costs associated with each person on a payroll including: supplies, hardware,

Technical Career Survival Handbook. http://dx.doi.org/10.1016/B978-0-12-809372-6.00076-1

materials, consultants, contractors, tests, travel, and software. Office space and furniture are generally considered overhead and often assigned as a percentage of the direct labor cost, that is, 20–30%.

Capital

Often manufacturers require various tools, patterns for castings, forging dies, and test stands that will be utilized for production over a period of 5, 10, 15 years, or until obsolete. These components are known as capital investments or fixed assets. Using the test stand as an example, there are several steps required prior to being operational. The stand must be designed (man-hours), fabricated (material and labor), installed (labor), and tested (labor). The collective cost of the capital investment is *depreciated* through a systematic reduction in the recorded cost of the fixed asset over its assumed useful life in accordance with the internal revenue guidelines. Therefore when the cost of producing a product is determined, the annual depreciated cost must be added to the manufacturing cost. A straight-line method is often used whereby the total capital cost is divided by the years of useful life to calculate the annual cost of the capital assets.

In a previous section, I listed goals associated with a business plan to successfully introduce canned motor pumps in to the US chemical processing market by third quarter 1987. Each goal had a budget and schedule. How is the budget determined? Each year most businesses must go through a planning process, often late in the year that pertains to the coming-year operations. During the planning process, equilibrium must be achieved between goals and their associated budgets versus what is anticipated for the business objective. Establishing a budget is an iterative process. At SFH, our business plan would focus on the coming year while also addressing the out years in order to include long-term projects (Fig. 76.1).

Often business objectives trickle down as a directive from the General Manager or the board of directors during the annual business planning process without realizing the resulting financial impact. Consider the CMP business plan objective: successfully introduce canned motor pumps in to the US chemical processing market by third quarter 1987. Since businesses operate with financial constraints, business plan objectives must be analyzed for fiscal feasibility. If the cost impact is unacceptable, the objective must be revised or eliminated. In the canned motor pump example above, perhaps the introduction should be pushed to the year-end 1987. This measure might reduce engineering, manufacturing, assembly, and test overtime and expedited shipment costs to an acceptable level. But would the delay be acceptable to management? What are the market consequences of delaying the introduction? These are the tradeoffs that must be considered by management.

After the business plan and the goals are established, the man-hours, expense, and capital estimates for each goal become the operating budgets. The goals are assigned target completion dates for the departments and/or personnel. For example, goal #5, sales literature for the CMP line, was developed using the Japanese literature as a guide while formatting the sales brochure consistent with other existing SFH product lines. Even the product name was derived from the use of an existing trademark, "Sundyne."

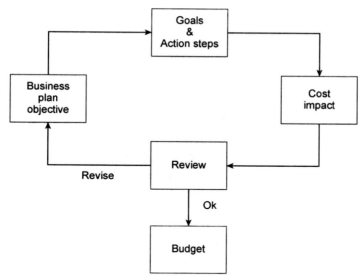

Figure 76.1   The budget process.

Song: "Time Erases Everything"
As popularized by: Burke
2008

Time Management

77

You may not be promoted to a manager position but you will definitely need to plan the way your time will be spent, hence time management. If a completion date is not assigned to your project you will likely be asked to provide one and make it known to your supervision. If your estimate is reasonable and accepted, you will be held to it.

During my first job as a design engineer, I was asked by the department head how long it was going to take to complete a shaft assembly design for the Boeing 737 ECS. In the interest of being cooperative, I pulled a number out of thin air that I later realized was hopelessly optimistic. I was so far off on that estimate that I have not forgotten the moment. Hopefully, the guidance in the following will keep you from falling into the same trap.

Here is my top-10 list of ways to manage time effectively:

1. **Establish the perimeter**. This in military term refers to understanding what is in the scope of the project and what is not. Be certain that the project manager and others are aware of the "what is not" in your scope of work.
2. **Break it down**. Define manageable parts (divide and conquer) so you can be sure you do not spend time working on out-of-scope tasks. In the example mentioned previously with the ECS, we reduced the design work to (1) rotating assembly, (2) compressor rotor, (3) compressor housing, (4) turbine rotor, (5) turbine housing, (6) fan rotor, and (7) fan housing. Each design task had an assigned man-hour target and was reviewed and approved separately.
3. **Plan**. Assign completion dates inside the overall project completion target dates. Put these shorter-term dates in your planner so you can see them every morning when you open your physical or virtual planner.
4. **Start**. The most difficult task with any project is beginning the work. Think of the Nike expression, "just do it." Unfortunately you will realize that by the time the project is assigned to you, you will already be behind on the overall program schedule, guaranteed.
5. **Tackle**. Do the tough stuff first and get it out of the way. You will feel more enthusiastic about the total project when you see what you have accomplished early on, knowing the rest will be a piece of cake.
6. **Look ahead**. You may need to order materials or software in advance due to their "lead time." This way you ensure whatever you will need will be ready later. The same is true for requesting assistance from a designer, technician, or metallurgist. They have commitments also and must manage their time as well.
7. **Avoid**. Stay away from those situations in the office that you are certain to be big time wasters. Do not allow vendors to provide a lengthy presentation on their capabilities and products when you know you will not require them for your current project.
8. **80–20**. Be aware that most technical personnel will spend 80% of their time on 20% of the task. Do not spin your wheels, you may need to ask for help if you hit a road block.

Technical Career Survival Handbook. http://dx.doi.org/10.1016/B978-0-12-809372-6.00077-3

9. **Develop contingency plans**. If you foresee a risk to the schedule, develop alternative actions. Similarly, if you anticipate finishing early, evaluate the consequences.
10. **Finish**. This is the second most difficult task on any project. Our human nature always makes us think we need to do just a little more to achieve perfection, make it look pretty and put a bow on it. I have known wise project managers to schedule design reviews well in advance so the date is cast in concrete for all involved to see from the outset serving as a target.

If you happened to be assigned as the project manager and develop a detailed schedule, be certain to review it carefully with your team so they can appreciate how their contribution is important to the survival of overall mission and that they buy off on it.

Song: "Time Is on My Side"
As popularized by: Rolling Stones
1963

Specifications

One of the basic "tools" engineers, scientists, and technicians prepare is a technical specification for materials, components, systems, or equipment. It is usually written to describe how something is to be developed or procured as if it had never been provided before. "Musts" are described specifically while "wants" are defined in a general manner. Collectively, they determine what the acceptable result should consist of. When they are completed and approved, specifications are often sent to suppliers of try to match their goods and services to those defined in the document. More on that process later.

In a previous section, I mentioned that I was contracted to Alstom Power, working on the La Paloma Generating Company, LLC, 1124 MW, natural gas-fired, combined cycle facility. I was responsible for preparing the specifications that described the construction requirements for the general contractor. By the time I completed the assignment, I had 17 three-ring binders of drawings, specifications, and standards representing our requirements to submit to various contractors for bid purposes. Many of the documents in the package were developed for other installations while some were prepared specific to the La Paloma plant. See Fig. 3.1.

Here is an example to illustrate the purpose of a specification. Old Dominion Electric Coop in Louisa VA requested that ITAC convert 6 and 10 inch #2 fuel oil storage tank valves from manual to electric so they could be operated from a remote station. This involved both mechanical and electrical disciplines to complete the project. After a site visit, my first significant task was to prepare a specification for the valve operators which included headings summarized as follows.

- *Scope of work*: This describes the design, purchase, delivery, and installation of the product, electric actuators, the location of the installation, and how they are to be furnished for the project; what work is to be done by the vendor and what will be done by "others."
- *Documentation*: The applicable reference documents shall be listed in this section. They may consist of codes, industry standards, and customer specific engineering standards such as painting, shipping, and marking.
- *Technical requirements*: This section describes how the actuator performs, its size, motor characteristics, gearbox, manual override, margins of safety, controls, electrical requirements and protective features, and visual operational displays.
- *Quality control*: The applicable quality requirements shall be defined such as the vendor's standards, applicable codes, or those listed in the specification. The vendor may be required to submit nonconformance reports with recommended dispositions for approval by the customer.
- *Examination and testing*: The requirements for vendor testing, inspections, and examinations are described based on applicable codes, standards, and the vendor's practices. Performance tests may be witnessed by the customer and curves submitted with the documentation package.

Technical Career Survival Handbook. http://dx.doi.org/10.1016/B978-0-12-809372-6.00078-5

- *Preparation for delivery*: This section should cover, cleaning, painting, storage, handling, nameplate data, packaging, warranty provisions, and filed service technician requirements.
- *Documents*: The customer's requirements may require submitting calculations, data sheets, schedule for manufacturing, electrical wiring diagrams, test data, inspection reports and installation manuals, dimensional drawings, and component weights.
- *Attachments*: These may include secondary specifications, drawings, and data sheets partially filled in, and other pertinent documents.

Often the buyer will prepare a general specification for, say, an actuator but attach a data sheet with project specific information in order to reduce man-hours for preparation prior to submitting the request to vendors.

A preferred way to reduce the man-hours to prepare a specification is to use a "go-by." This entails simply using a previously prepared specification as a text template, deleting the nonpertinent requirements, and inputting the pertinent requirements and data. Consultants typically have standard specifications and data sheets for equipment that serve as templates for developing new specifications rather than starting from scratch.

Song: "Handy Man"
As popularized by: Jimmy Jones
1960

Documents

Engineers, scientists, and technicians communicate technical information and data by formal, approved means through many different types of documents using hard copy and electronic media. These tools ultimately end up stored in electronic media and the hard copies are destroyed. Here is a list of the most common documents used by technical personnel.

Drawings: They show the physical arrangement with pertinent dimensions, some of which may be for reference and not construction or manufacturing. For part drawings, tolerances and finishes are included along with reference specifications such as painting, inspection, and quality control.

Marked drawings or sketches: This may seem redundant but drawings that are physically marked up to show changes are an important tool. They are used to show modifications to parts, structures, systems, and circuits so that the impact of the changes may be evaluated before the final production configuration is determined.

Bill of Materials (BOM): These list part numbers, part names, descriptions, quantities for each assembly, and often vendor part numbers. A BOM is prepared for a one-off assembly and often used as a basis for estimating the cost.

Equipment Lists: These include lists of multiple manufactured assemblies, vendor purchased assemblies, and quantities of parts.

Process Flow Diagram (PFD): These are the basic schematics that define a system in a block diagram format with components, piping, and valves and includes operating data such as design flow rates, pressures, and temperatures. It is the first diagram produced from which the plant and instrument diagram and other documents are developed.

Plant and Instrument Diagram (P&ID): These diagrams are produced including mechanical, electrical, and chemical engineering information. Schematically they show flow direction, piping sizes, valves open or closed, controls instrumentation, vessels, nozzles and sizes, heat tracing, and equipment and instrument numbers.

Calculations: These are developed under all disciplines and include the problem definition, knowns and unknowns, assumptions, formulas and methods used, the calculation, results, and conclusion. Calculations are normally peer-checked for accuracy and approved.

Inspection reports: These may be generated by field service to report the condition of equipment, components, or materials installed in the field and include physical condition, damage assessment, operating and performance data, and recommended changes.

Electrical Schematics: Sometimes called wiring diagrams, these are produced by the electrical engineers schematically defining how electrical components are interconnected. They include symbols, terminals, indicators, contactors, and circuit-protection devices for a particular device.

One-Line Diagrams: These are produced by the electrical engineering department to represent a three-phase power system in block diagram format. Included as symbols are circuit breakers, transformers, capacitors, bus bars, and conductor (wire) sizes.

Technical Career Survival Handbook. http://dx.doi.org/10.1016/B978-0-12-809372-6.00079-7

Instruction Manuals: These are developed by most disciplines and include the product description, installation, operation, maintenance, trouble shooting and diagnostics, safety warnings, recommended spare parts, and attachments.

Work Order: Theses are developed by all disciplines and generally written to define work to be done by construction, service, or an inspection organization. They include scope of work (SOW), demolition, new construction, inspections, materials required, applicable specifications, codes, and drawings.

Capital Authorization Request (*CAR*): When monies are to be appropriated for investments in tooling, software, or machinery, this document is utilized to obtain the necessary approvals. It should include the name of the initiator, amount, timing, purpose of the investment, economic justification, and the required approval signatures.

Preparation of these documents may not sound very glamorous but they represent the necessary, challenging, nuts-and-bolts work performed by many technical personnel. All the documents are usually peer-checked and approved by supervision prior to being released for the development of a product or project. Consulting engineers typically submit these documents to their clients for approval and/or for general information often including a professional engineer's stamp.

Song: "Book of Love"
As popularized by: Monotones
1958

Performance Appraisals

This activity, although very important in the development of technical personnel, is often dreaded by both employee and supervisor. To a degree, an employee's performance should receive comment from supervision practical daily so that there are no surprises when a face-to-face performance review is conducted semiannually or annually. I have participated in several different appraisal formats which are described in the following.

Performance Review of Goals and Objectives

This approach requires that a list of performance objectives for the subordinate are established and approved by the supervisor at the outset of the review period. The objectives may be extracted from the business plan and/or personal development goals such as completing a particular course or seminar. In either case the objectives must be specific and measurable. At the designated review time, the results for each objective are assessed by the supervisor and recorded. The assessment may be used as a basis for adjusting salary.

Performance Review of the Job Description

The subordinate must have a known, written, and approved job description in advance of the review date in order that he/she is working under it throughout the review period. Thus when the review is conducted by the supervisor, their comments will be either positive or negative in conformance to the job description. This often results in vagaries and opinions more so than the review of performance to goals and objectives approach.

Technical Career Survival Handbook. http://dx.doi.org/10.1016/B978-0-12-809372-6.00080-3

Review of Performance Characteristics

In this format, specific job dimensions must be selected which are relevant to the subordinate's position. I was evaluated under this method using job dimensions such as planning, leading, controlling, organizing, cost consciousness, creativity, judgment, and cooperation. The supervisor prepared written responses to each dimension and evaluated them on a scale from 1 (marginal) to 5 (superior). The characteristics can also be weighted. By calculating the average score, a quantitative result is determined and can be related to a salary adjustment.

Co-reviews

I have experienced this procedure working at two different companies. It requires that separately, the subordinate and the supervisor complete performance reviews. In the case of the former, he/she must comment on their assessment of their performance to job dimensions or objectives. The latter also reviews the subordinate's performance and makes specific comments on strengths, weaknesses, and development needs. With both reviews on the table, the performance review/discussion requires agreement by both parties going forward. This approach may be awkward particularly in the case of substandard employee.

In all the above methods, there should be a final section for comments by both the supervisor and subordinate. This is an appropriate section to comment on training required, possible promotions, or even transfers. Regardless of the appraisal method employed, communication is the key. Ultimately, the performance appraisal gets filed in the employee's permanent records and each employee should retain a copy for their personal records along with any salary or position level changes that happen or disciplinary actions that follow.

Contrary to what has been said in the preceding paragraphs, since 2006 or so, some companies have tried very hard to maintain generic rather than specific job descriptions, and some have decided not to write job descriptions at all! By so doing, employers have huge leverage to treat employees as if they are working under a personal services agreement. Often this works to the disadvantage to the employee. In a related way, the term hired or working "at-will" means that you can be fired without good cause or "for any reason." These are indications that your employer follows a policy which may be a detriment to your survival. You may want to seek advice from an attorney if you are required to sign an employment agreement containing this language.

Song: "Every Breath You Take"
As popularized by: The Police
1983

Decision-Making

In a previous Chapter 36, I mentioned a type of training that encompasses various subjects as determined by the management of the company, which I refer to as elective specialties training. One particular course I completed in this classification was decision-making. I did not realize at the time that the principles behind the course would stay with me for decades as a very useful tool.

The course was based on principles derived by a company called Kepner-Tregoe, Inc. (KT) who documented their methods in a book entitled *The Rational Manager*. The authors saw that many important business decisions were made in a reckless manner with costly consequences. They researched work that had been done in this field and dissected the thinking involved in each step of the process. One particular method they developed, I refer to as the "musts–wants technique."

In a previous section, I discussed the dilemma of the notorious counter offer that may occur when you decide to quit your job and accept another job elsewhere. Then, unexpectedly your current employer offers you a counter deal. What do you do? The KT method will be helpful in resolving this dilemma.

Also in a previous section, I discussed the decision to market the CMP product line of seal less pumps under a joint venture with the Nikkiso Co. in Japan. The alternative was to design, develop, and manufacture a comparable product in our plant in Colorado. In a greatly simplified example, using the KT analysis, the following steps would have been necessary to reach a decision.

Step 1. Consider the alternatives
    Alt. 1—Do nothing
    Alt. 2—Market the CMP in the US as manufactured in Japan under a JV
    Alt. 3—Manufacture the CMP in Colorado under a JV with Nikkiso drawings.
Step 2. Determine the *musts*
    M1—Introduce a CMP product line
    M2—Have 12 units running successfully in the field in 8 months
    M3—Utilize existing personnel.
Step 3. Determine the *wants* and assign weights 1–10 as most important
    W1—Minimize risk of implementation 10
    W2—Ease of implementation 8
    W3—Minimize capital investment 4
    W4—Use existing sales network 9
    W5—Minimize expenses 6.
Step 4. Evaluate musts
Step 5. Evaluate/score wants from 1 to 10 as most favorable
Step 6. Select best score in Table 81.1.

Technical Career Survival Handbook. http://dx.doi.org/10.1016/B978-0-12-809372-6.00081-5

Table 81.1 **KT Analysis Example**

Musts	Alt. 1	Alt. 2	Alt. 3
Introduce a CMP product line	No	Yes	Yes
Have 12 units operating in field in 8 months	No	Yes	Yes
Utilize existing personnel	No	Yes	Yes
Wants			
Risk 10	No	$10 \times 9 = 90$	$10 \times 5 = 50$
Ease 8		$8 \times 7 = 56$	$8 \times 5 = 40$
Capital 4		$4 \times 10 = 40$	$4 \times 5 = 20$
Sales 9		$9 \times 10 = 90$	$9 \times 10 = 90$
Expenses 6		$6 \times 8 = 48$	$6 \times 1 = 6$
Score	0	324	206

The results of this simplified problem eliminate Alt. 1 as it does not satisfy the musts where Alt. 1 and Alt. 2 do satisfy the musts. However, Alt. 2 is selected due to its higher wants score.

Sometimes the result of this analysis will be obvious during the process. What is the major advantage of this method? When completed, it serves as a documented record of the basis for the derived decision and *all those that participated* in the process. Assigning weight and scores will make for an interesting negotiation process in a group session.

Song: "It's Up to You"
As popularized by: Ricky Nelson
1963

Project Team

82

In a previous section, I mentioned an organization structure that requires a project manager, program manager, or task force leader assigned temporarily to lead a project. Let us look at this structure more closely.

First—Why Have a Temporary Organization?

In a consulting engineering company, it is very common to assign various discipline engineers and designers to focus sharply on a client's design needs. The group will remain together as a project organization but may be also committed to other projects outside the scope of the temporary assignment. Typically, the team would not be temporarily located in the same office/room. In a manufacturing organization, frequently a design team is temporarily established to develop a new product or product feature wherein they are all focused on a single product objective. While I was used by SFH, we would assemble a task force in reaction to a field failure/problem. This would often require designing a modification to an existing product, ordering material, manufacturing a component, shipping it to the field, and completing a retrofit. All these tasks would be accomplished on a highly expedited basis.

Second—Who Should Participate on the Team?

All disciplines necessary to complete the objectives within the required time frame. They should be considered as coequals. In the case of a problem with a product installed in the field, a field service member should be included on the team to coordinate activities at the jobsite. In a previous section, I mentioned going to work for

Technical Career Survival Handbook. http://dx.doi.org/10.1016/B978-0-12-809372-6.00082-7

the Dominion project manager assigned to the power plant to be constructed along the Potomac River. I was part of a multidiscipline team, which I used to refer to as the "dirty dozen" based on the movie. The team consisted of mechanical, civil, electrical engineers plus safety, scheduling, quality control, and inspection technicians with the mission of supervising the plant construction. It was a memorable experience (Fig. 50.1 of Chapter 50).

Third—What Are the Qualifications of a Project Manager?

This person must have a keen understanding of the problem or project as well as good relations with the customer. Also, it is important that they understand each team member's capabilities and a reasonable expectation of their ability to complete the work. The project manager may be of whatever discipline necessary to be familiar with the project objectives. Good communication and organizational skills are a must.

Fourth—How Is the Workload Organized?

The objectives must be clear and specific to all task force members. A schedule should be prepared defining all the required goals, completion dates, and responsible personnel. This completed schedule should be copied to all members of the team for review and comment.

Fifth—How Is Group Progress Communicated?

Weekly reports should be prepared, circulated to all members of the team as well as the client and the GM, and discussed during weekly meetings. Monday morning is always an ideal time to review the status of projects. Also a list of roadblocks that have occurred should be maintained and the action plan to overcome them.

At the conclusion of the project, the discipline managers should collaborate with the project manager to assess the job performance of the team members for reflection in their annual performance reviews.

Song: "With a Little Help from My Friends"
As popularized by: The Beatles
1967

Scope of Work

83

Whether you are required to develop a new epoxy, a pump, a circuit, a boiler, an e-cig, or a special purpose building for a client or your boss, you will need to define the scope of work (SOW) in a written document. This will enable you or others to establish a time frame for scheduling the work and a budget to finance the effort and clearly state what the deliverables (documents) are required. Frequently, SOWs are prepared and determined to be too expensive or long in duration to be acceptable, then often they are revised. The SOW also pinpoints the work so that "scope creep" is avoided or minimized. Often this occurs when an unanticipated roadblock or problem occurs during the project and a dispute arises as to whether it is covered under the project SOW.

To illustrate a SOW document, let us take a simple example based on aging equipment in a chemical process plant. Here are the main headings and subheadings that should be addressed:

1. **Project objective**—The cooling tower and chemical addition equipment in Area 9 have been in operation for 40 years. The chemical addition skid is a constant source of maintenance and operational problems. The objective of this project is to prepare a conceptual engineering study to evaluate/compare the current situation with lower cost and more reliable alternatives, which will entail replacement and/or modifications to the existing of the chemical equipment skid.

2. **Assumptions and clarifications**—The chemical addition equipment in Area 9 has developed leaks, causing ground contamination, clean up, and excessive maintenance. The chemical pumps are unreliable, do not maintain calibration, and require frequent inspection and adjustments. The client has agreed to provide reliability data, arrangement drawings, energy costs, and estimated cost of failures. No specific deficiencies in performance or reliability have been defined by the client but the existing conditions are unacceptable.

3. **Design criteria**—The client will provide existing P&ID and flow diagrams. Since chemical skid power consumption is negligible, electricity usage will not be a design consideration. No engineering specifications will be prepared for this study. Preliminary vendor proposals will be evaluated for applicability for this application.

4. **Project requirements**—This project is a conceptual study and does not entail design, construction, or demolition work. Options for modification or replacement of the equipment, estimating piping modifications, electrical controls, wiring, and structural supports will be evaluated.

5. **Applicable codes and standards**—This study shall consider applicable organizations, standards, codes, state, local and federal rules and regulations, and any special requirements by the client. Governing codes may include but not limited to ANSI, ASME, NEC, NEMA, and OSHA.

6. **Client specifications**—This will be considered where applicable.

Technical Career Survival Handbook. http://dx.doi.org/10.1016/B978-0-12-809372-6.00083-9

7. **Area classification**—Nonhazardous or general purpose.
8. **Safety and environmental concerns**—This study will assume that the current safety standards and procedures shall be applicable to any configurations proposed. The study will assume that double wall containment with interstitial leak detection and alarm for the chemical addition skid will be furnished.
9. **Maintenance concerns**—Personnel access for maintenance and inspection shall be provided for the equipment skid proposed under this study.
10. **Engineering design requirements**—Equipment shall be generally described but not sufficiently specific for ordering purposes. Tie-in piping points and any changes in utilities identified. A description of demolition, if any, shall be provided sufficient for cost estimating.
11. **Drawings and software deliverables**—Engineering report, schematics, photos, sketches, and equipment layouts shall be provided as required. No software will be provided to the client.
12. **Quality assurance**—A peer review of the results and recommendations of this study will be conducted.
13. **Schedule**—The completion of this study will be in 6 weeks of the acceptance of this SOW by the client.
14. **Construction**—No construction work will be authorized under this study.
15. **Risks**—Costs presented in this study will be best estimates and subject to change and escalation in the future.

Based on a mutual consultant–client agreement of this SOW, an estimated project cost may be determined. Often a proposal is prepared as the next step, which will include the agreed upon SOW, the lump sum price, terms, and conditions under which the work would proceed.

Song: "The Gambler"
As popularized by: Kenney Rogers
1978

Procurement

84

Frequently, engineers, scientists, and technicians will participate in the procurement of equipment, materials, and services for in-house use or for clients. In fact, service companies and consultants do not manufacture any products. Therefore, they must take several steps to ensure that the correct goods are purchased. Some items may be the vendor's standard while others may be engineered for the specified application.

In a previous section, I described preparing a voluminous specification for the construction services and received response bids ranging from $28MM to $175MM. While of course we preferred University Marelich Mechanical's (not really a "university") low bid, our concern was that they did not fully understand the scope of work (SOW) while the higher bidders probably did not really need/want the work. You must be cautious enlisting the services of a contractor who does not understand the SOW to ensure they do not go bankrupt half way through the project. So we worked with UMM to educate them on the scope of the project and eventually they arrived at a revised bid of $64MM and we accepted their offer. Often a performance bond, aka a contract bond, is issued by an insurance company or a bank to guarantee satisfactory completion of a project by the contractor. Also, a job requiring a payment and performance bond will usually require a bid bond, just to submit a bid for the project.

Here are a few of the steps normally required in the procurement process to ensure a favorable outcome (Table 84.1).

Complex components, equipment, and skid may necessitate a visit to the vendor's site to witness construction, testing, and inspection. I specified equipment skids while working for Adtechs requiring me to be on site at the vendor's fabrication facility for several weeks while the skids were being assembled. This was necessary to ensure that

Table 84.1 Procurement Process

Activity	Responsibility
Prepare and approve the specification	Engineering
Submit a request for quote (RFQ)	Engineering
Obtain vendor/contractor quote	Procurement
Evaluate vendor/contractor bid	Engineering
Revise specification if required	Engineering
Prepare a requisition	Engineering/project engineering
Issue purchase order	Procurement
Manufacture product/equipment/skid	Vendor/contractor
Receipt inspection	Quality control

Technical Career Survival Handbook. http://dx.doi.org/10.1016/B978-0-12-809372-6.00084-0

when completed and shipped to the site, there were minimal discrepancies and rework. Field rework/changes are always certain to be pricey.

Song: "Can't Buy Me Love"
As popularized by: The Beatles
1964

Trade Shows

85

Trade shows are a common medium by which manufacturers and engineering companies present their capabilities to prospective clients. There are hundreds of trade shows held annually throughout the United States and the world covering a broad array of fields. A few of some of the largest shows and a small sample of the subjects are shown in Table 85.1.

While I employed by SFH, we participated in several trade shows each year using the opportunity to focus on new products, product features, and presentation of technical papers. When a company presents a product at the show, they are also revealing some of the technology behind the product, which creates a dilemma with competition who will also likely to be visiting the booth. So a judgment call must be made as to what is most beneficial to display and describe to the public.

Table 85.1 Major Trade Shows

Trade Show Name	Location	Subjects
Plant Maintenance, Inspection and Engineering Society	Pasadena Convention Ctr. Pasadena, CA	Rotating equipment, engineering design, welding, electric motors
Pump and Turbomachinery Symposia and Expo.	Brown Conv. Ctr. Houston, TX	Steam and gas turbines, compressors, expanders, and pumps
Consumer Electronics Show CES	Las Vegas Conv. Ctr. Las Vegas, NV	Electronics, training, education, engineering standards
World Energy Engineering Congress, WEEC	Orange County Conv. Ctr. Orange County, CA	Energy management, power equipment, building automation, energy controls
GlobalCon Conference and Expo.	Pennsylvania Convention Ctr. Philadelphia, PA	Energy management, renewable and alternative energy, plant facilities
International Manufacturing and Technology Show, IMTS	McCormick Place Chicago, IL	Machine tools, robotics, automation, electrical systems
Power-Gen	Orange County Conv. Ctr. Orange County, CA	Cogeneration, fuel suppliers, electrical power, construction contractors

Adapted from events in America 2014 data.

Technical Career Survival Handbook. http://dx.doi.org/10.1016/B978-0-12-809372-6.00085-2

Most importantly, there is a tremendous opportunity to reach new customers during a trade show and possibly discover new market applications. The booth should be manned with engineers, scientists, technicians and sales engineering personnel, whoever is capable of addressing any technical questions that might arise, on the spot. Usually, prospective clients are urged to complete a *lead* form identifying them by name, address, company, and area of interest. Following the show, the leads are distributed throughout the sales organization for follow-up with the potential customer. In some instances, a current user of our products would appear and provide an opportunity to gain feedback on a product performance in the field.

The aspect of a trade show that I looked forward to the most was visiting competitive product booths. I was always fascinated how different product designs could produce similar performance. It was also interesting to note the nuances between ours and their products and the products and the competition were emphasizing. I would often take literature they were distributing, review it in detail, and pass on my comments to my company management. Also different ideas for products or product features may result from attending a trade show.

Another aspect of trade shows is an opportunity to visit with vendors that may be current or future suppliers of components or materials you need to incorporate into your product. For example, we used many different motor manufacturers in conjunction with our pump lines at SFH, i.e., direct current, alternating current, and variable speed in different hazard classifications. Manufacturers of these products were usually present at trade shows.

Finally, trade shows offer an opportunity to network with other engineers, scientists, and technicians in your industry who you may want to contact in the future. Some of these acquaintances may be important for your career survival. Many may be members of societies in which you are a member. Also, trade shows are often attended by consultants that may be able to assist you in the future with the design or marketing of your product.

Song: "Island in the Sun"
As popularized by: Weezer
2001

Vendors

86

Many products consist of a combination of made parts or purchased parts. Equipment manufacturers often must decide if they will machine a component or find the part through supplier. This is known as a "make or buy" decision. In most cases, it is decided based on quantities cost and availability. At SFH, we had extensive machining capability so we often leaned toward producing parts in our own shop even when a vendor could produce the part for a lower cost. In a chicken and egg approach, it was thought that if we were too aggressive in buying on the outside, we would be underutilizing our machinery and be dependent on vendors for deliveries. Then, we would not have machinery and machinists available to produce the parts that we were not able to procure through vendors. So we were bullish on making parts in-house. Obviously, standard seals and fasteners would be "buyouts."

Good vendor relationships are important in assuring timely assembly and delivery of complex equipment. It is wise to maintain a preferred vendor list and a history of their ability to meet delivery dates. Occasionally, an expedite fee is required by a vendor to speed up delivery of a component to cover air freight and overtime charges.

Many companies today have replaced the purchasing or procurement department with what is referred to now as "supply chain management" where a majority of a manufactured products might be made from vendor-supplied parts worldwide. This is particularly true in the aerospace industry.

Engineering consulting companies rely on vendors to submit pricing for equipment and materials in compliance with specifications and drawings. Often, the supplier's price and delivery information is incorporated into a client proposal, which may or may not go forward pending the client's decision. This will require that the vendor invests time to prepare a proposal knowing that it may not become a reality. So accuracy and prompt response are additional qualities of a preferred vendor. Often alliances or special blanket pricing are negotiated with vendors based on annual quantities of purchase products. In a previous section, I mentioned an agreement whereby products were purchased from a manufacturer of air filters and dryers then rebranded and sold under a different manufacturer's name.

In recent years, I have seen a great deal of emphasis on vendor presentations in the workplace. In addition to product literature, vendors that offer products such as valves, controls, fluid conditioning, and chemical injection systems are often willing to conduct technical training seminars in conjunction with their products. Some offer continuing education credit hours applicable to professional engineering requirements. You may be surprised to learn that some vendors can be important allies throughout your career.

Song: "For Those about to Rock"
As popularized by: AC/DC
1981

Technical Career Survival Handbook. http://dx.doi.org/10.1016/B978-0-12-809372-6.00086-4

Meetings

87

Webster's dictionary defines meetings as "a coming together of persons or things for a common purpose." They are a normal part of the day for engineers, scientists, and technicians and are used to disseminate or collect information. Here are a few types of meetings you are likely to encounter the following:

- Staff meetings—These are normally top-down, supervisor to subordinate information sessions. I have participated in many of these departmental, unidiscipline meetings usually conducted on Monday mornings covering the events of the previous week and what lies ahead for the coming week. A monthly version is also common, whereby the month's activities and financial data are presented and scrutinized in detail. These meetings are usually conducted by going around the table taking turns at discussing what the subordinates think are important issues to share with supervision and others.
- Design project/product review meetings—These are usually multidiscipline meetings attended by all those with some involvement or who are affected by the project or product activity so as to obtain their reaction to change. For example, when a new product is on the verge of introduction, what new tooling will be required in the shop for manufacturing purposes? What software will be necessary to evaluate product applications? What safety precautions will be mandatory on the project job site? How will the new project work load impact the current staffing level?
- Progress meetings—These are often called status report meetings and are normally multi discipline and attended by all those with some involvement with the project or product activity. The purpose is to evaluate the progress toward established goals and milestones. In the case of a field problem, for example, have the replacement parts been purchased, on hand or shipped to the field for installation? Has a failure analysis been conducted on the failed parts? Was the quality control inspection completed on the replacement parts? Do we have manpower on site for the retrofit program? Do we need special inspection services on site?
- Brainstorming meetings—These are likely the more creative meetings were no holds are barred for the purpose of arriving at the best path forward. They are attended by multi- or unidiscipline personnel selected as knowledgeable of the product or project and most likely to volunteer solutions. Perhaps it is a senior, experienced product engineer, or even the recent graduate that is extremely computer savvy. Depending on the subject, an outside consultant may be called in to offer suggestions even though they may not be intimately involved in the implementation. Crazy suggestions should be welcome here so no holding back. It is best to start making a list of any and all ideas and evaluating them later. Then, the previously described KT problem-solving method can be applied to weigh the "musts and wants" effectively.

While working at Worthington Pump Corp., we experienced a problem with vertical flood water pumps installed in Albuquerque NM that were required to operate only in the event of a major rainstorm. During other times, they were dormant but subjected to fouling in the shaft bearings due to the presence of nasty road dust.

Technical Career Survival Handbook. http://dx.doi.org/10.1016/B978-0-12-809372-6.00087-6

Figure 87.1 Bearing lubrication system retrofit.

So when the pumps were energized in an emergency situation, they repeatedly failed to run. During a brainstorming meeting, we arrived at a "field fix" solution to retrofit the pumps (Fig. 87.1). An automatic lubrication injection system was provided that would purge the cutlass rubber shaft bearings (normally lubricated by the flood water) periodically with grease serving as a barrier to the dust. So when a flooding condition occurred, the pumps would operate reliably. They did and we received our delayed final payment for the equipment.

Regardless of the type of meeting conducted, there are several simple guidelines to ensure that the meeting will be productive and you survive the mission. I call them the **R.A.P.I.D.** approach.

> **R**eason and objectives for the meeting should be published and distributed to attendees well in advance so they can prepare. The agenda should include results/minutes from the last meeting on the same project/mission.
>
> **A**ction items based on consensus from previous meetings as well as new items should be documented. Should additional personnel be selected to supervise the installation? Should spare parts be shipped to the site as a precaution?
>
> **P**rogress achieved from previous meeting action items should be discussed and evaluated. Were materials shipped to the field? Was the assembly of the equipment completed? Any setbacks?
>
> **I**ndividual's responsiblity should be identified and published in the minutes. Backups may also be listed if necessary, including contact information to reduce risk of failure of the mission.

Dates for actions to be completed should be targeted and published in the minutes to the meeting. The actions should include the schedule for future review meetings to be conducted to provide project direction at the effective intervals.

If a decision-making process like KT is conducted during a meeting, it is wise to attach the factors and results to the meeting minutes. This will document the logic utilized during the decision-making process.

Song: "Minute by Minute"
As popularized by: The Doobie Brothers
1978

Transfers

There are numerous reasons for both voluntary and required transfers within a company. You will have to be the judge of whether a particular transfer is in your best interest. Listed are a few factors to consider.

1. Does the transfer provide an opportunity to learn a new or enhance an existing specialty within your discipline?
2. Does the transfer offer a better promotion opportunity in future?
3. Does the transfer offer an opportunity to work under supervisor considered to be a rising star in the organization?
4. Does the transfer offer an opportunity to relocate to a more desirable location?
5. Does the transfer enable you to work in conjunction with new or more challenging technologies?
6. Is the transfer a short- or long-term situation? If short term, is it worth the disruption in your career path?

My experience with transfers was limited to one voluntary situation while at GE Medium Gas Turbine Division. GE had just hired a new department manager to supervise a combined cycle design and application group. Leroy Tomlinson was a well-known author, inventor, and consultant. He had vast experience in gas turbine and combined cycle system design and application, gas turbine and combined cycle electrical power plant design, and design and application of mechanical-driven gas turbine systems for pipeline and process compressors and oilfield gas recompression. At that time, combined cycles were expected to result in higher efficiency operation resulting in increased power generation and lower fuel consumption. This was an advantage that was certainly desirable in the power plant market subsequent to the Arab embargo event in the 1970s.

I was feeling a bit stagnate in my position as a controls components design engineer so I decided to introduce myself to the new manager shortly after he arrived on the scene. Soon I learned that he would be in need of an application engineer to prepare combined cycle proposals. So I approached my current supervisor to seek permission to formally interview for the position and if accepted, apply for an interdepartmental transfer. He did approved and after finishing a few current projects, I completed the transfer.

In retrospect, my transfer accomplished several things. Considering the factors listed earlier, it was an opportunity to learn a new specialty, I got in on the "ground floor" of a new department, the manager was considered a good mentor, and I was learning a new and challenging technology. Because the position was newly established, I had incorrectly assumed my transfer was a lateral move. So I was pleasantly surprised when I was informed that the increased responsibility resulted in a corresponding increase salary class and therefore I received a raise.

Technical Career Survival Handbook. http://dx.doi.org/10.1016/B978-0-12-809372-6.00088-8

Contrary to my situation, companies often *require* personnel to transfer to a different department, or business unit for various reasons, one of which is to balance out the workload. In a previous section, I explained why at SFH, the company was reorganized into three product groups. Consequently, there were numerous transfers involved including my change from engineering director to director of Sundyne products. My mandatory transfer resulted in a promotion on the basis that I would be responsible for multiple disciplines. However, I suspected there were several other accompanying promotions made to encourage personnel to cooperate with the radical reorganization plan. Geographical transfers were mandatory for many of the sales personnel causing some to leave the company voluntarily or relocate to a different sales office to survive the reorganization.

Song: "The Letter"
As popularized by: The Boxtops
1967

Feasibility Studies

One of the more challenging activities involved in a technical position is preparing a feasibility study. Whether it is associated with a product, process, or a project, the study usually starts with a clean sheet of paper and a few preconceived notions. The purpose of a feasibility study is to provide direction as to which design or alternative is preferred based on criteria such as cost, projected sales, complexity, safety, reliability, and ease of implementing. Therefore the study needs *only be in sufficient detail* to arrive at the preferred alternative. Often a brainstorming meeting is a good way to kick off a feasibility study thereby getting input from others who may not be directly responsible for the study outcome but may have some applicable "off the wall" ideas.

To further illustrate the concept of a feasibility study, I am reminded of study that I completed which focused on a Wildlife Management Area (WMA) in Virginia. I divided the study in to seven parts and summarize them briefly as follows:

1. Executive summary—The purpose of the study was to determine the preferred concept for a permanent pumping station considering installed cost, energy cost, safety and security, and operability.
2. Current situation—In the Spring, water must be pumped from the wetlands area to the other side of a coffer dam into a river to allow fauna to grow. If natural rainfall does not reinundate the area, river water must be pumped into the wetlands usually during the Fall. The fauna growth provides forage for migratory birds when the area is flooded during the winter.
3. Long-term solution criteria—The desired solution should be an installation that is permanent, easily installed, operable automatically/unattended, secure from vandalism, nonpolluting, well lighted, reliable, low cost, and efficient to operate.
4. Design considerations—Three power sources were considered: solar, diesel, and electric. Performance of single and multiple vertical and self-priming pumps was evaluated. Other factors such as security fences, concrete buildings, containment of spills, steel versus PVC piping, costs, codes, and standards were reviewed.
5. Proposed installations—Five upgrade configurations were defined according to the type of pump(s), valves, drivers, mounting, fencing, and buildings required. Decision analysis, musts–wants, was used to evaluate and compare the concepts (Fig. 89.1).
6. Cost evaluation—Major cost components for each of the five upgrade configurations were estimated and summarized to obtain the installed cost. The 5-year operational energy cost was added to the installed cost to obtain a total 5-year cost estimate based on assumed operating hours, fuel, and electricity costs.
7. Recommended solution—Based on installed cost, energy cost, security, and operability, the recommended concept was a single electric motor-driven horizontal pump housed in a cargo (sea) van, with valves, piping, and strainers to flow water to either the river or wetland (Fig. 89.2).

The KT decision analysis was utilized in deriving the optimum configuration and the results were included in the final report.

Technical Career Survival Handbook. http://dx.doi.org/10.1016/B978-0-12-809372-6.00089-X

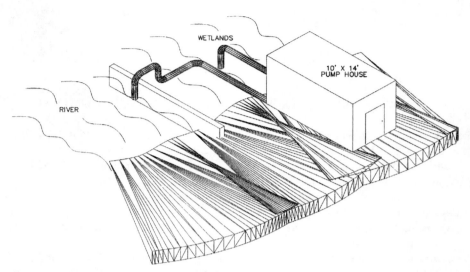

Figure 89.1 Pump house installation concept.

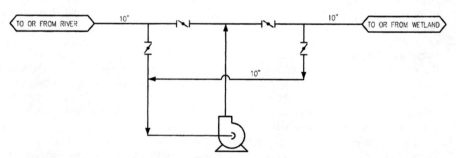

Figure 89.2 Wetlands pump flow diagram.

Feasibility studies may extend beyond the paper stage. A product or process may be feasible but not practical. For example, the oil-free air compressor I mentioned previously developed at SFH was determined to be feasible during the design and test phases. However, as the product was initially manufactured on a small quantity basis, cost and price projections were determined by the marketing department. As a result, after several field trial units were placed in operation at customer sites, the product was canceled from production after it was determined to be too costly to produce.

Song: "Waiting in Vain"
As popularized by: Bob Marley
1977

Projects

In previous sections, some of the documents "tools" commonly created and used in the technical world were described. Now let us look at an example of the many events listed in Table 90.1 that must occur during a typical project such as the manufacture of the chemical equipment skid shown in Fig. 4.1.

In this example, it is likely that four to six technical personnel would be required to complete the design work in a reasonably time period, perhaps 3–4 months. More personnel might be required or providing overtime to "compress" the schedule, if required. Additionally, on the manufacturing side, a multitude of fabrication, welding, machinists, assemblers, and inspectors would be required to produce the necessary components and assemble the skid.

More complex assembles such as medium gas turbines manufactured at General Electric in Schenectady, NY while I was employed there, required 52 weeks from receipt of order to ship date. The "backlog" includes the units that remain on the books to be shipped. This has a major influence on the ability of a manufacturer to meet "promised" ship dates, a positive for any company's reputation.

Lewis DelVecchio, a friend of mine, is an experienced engineering project manager who has conducted projects for clients such as Honeywell International, DuPont, Philip Morris, and other major corporations. I asked him what are the main reasons that projects might go poorly? Here are his suggestions.

1. Schedule miss

   The key to any successful project is a detailed milestone schedule with task durations and predecessors identified. This enables the project manager to identify the critical path items and closely manage them. It is important not to "over promise." Also, some "float" (additional time) should be included in the schedule. Why? Because factors like weather and client requested changes (change orders), which may not be within your immediate control. That is why scheduling tools are important for forecasting completion dates as well as feasibly possible. For small projects, Microsoft project is easy to learn and works fine. More complex projects are better suited to Primavera (see the previous section addressing schedules).

2. Cost increases

   Missing schedule dates is one thing but being over budget can have an adverse impact on a project manager's survival. Start with a detailed scope of work (SOW), and a bill of materials (BOM) that includes cost and delivery date data. This enables development of a detailed cost estimate including labor, engineering, and material costs. Always include a contingency of 5–10% or greater if the SOW is not well defined. It is always understood that the project/product must meet performance objectives, therefore design changes are inevitable. Evaluating scope changes and preparing change orders are essential for survival. Before any change is approved, the cost and schedule impact must be determined.

Technical Career Survival Handbook. http://dx.doi.org/10.1016/B978-0-12-809372-6.00090-6

Table 90.1 **Typical Project Steps**

Step	Activity	Responsibility
1	Complete the scope of work	Project manager
2	Complete the work schedule	Project manager or engineer
3	Assemble the team	Project manager
4	Establish the budget	Engineering and project manager
5	Kickoff meeting	Project manager and all involved
6	Complete 30% design	Engineering
7	Specify and requisition long lead equipment/items	Engineering
8	Order long lead equipment/items	Procurement
9	Complete 100% design, bill of material, and review	Project manager and all involved
10	Order all remaining components	Procurement
11	Supervise equipment vendor manufacturing	Engineering
12	Receipt inspection of materials	Quality control
13	Manufacture, assemble, and ship skid	Manufacturing
14	Supervise skid field installation	Engineering
15	System commissioning and test	Field service technicians
16	Final inspection	Engineering and QC
17	Issue closeout documents to client	Engineering

Song: "All Star"
As popularized by: Smash Mouth
1999

Promotions

91

There are several circumstances under which a technical employee is promoted to a higher or greater position of authority according to company job descriptions and salary classifications. While promotions are at the discretion of company management, they are not arbitrary and generally do not occur on a scheduled basis. In fact, this may not occur particularly if the employee's performance does not show significant improvement. Some of these promotion situations are as follows:

- *New business unit created*—This may cause the company to establish several new management and supervisory positions to absorb the anticipated workload and facilitate staffing.
- *Reorganization*—As I described in an earlier section, often multiple discipline responsibilities may require personnel to be placed in new management or supervisory roles.
- *Fill a vacancy*—Losing an employee due to a quit, promotion, or demotion situation will frequently result in an opportunity for an existing employee to move up.
- *Topped out in the salary class*—If an employee reaches the top of their salary class, management *may* consider moving them to an alternate position at a higher salary class or risk losing them via resignation if no action is taking.
- *Transfer*—As I described in a previous section, a move to another department within the same company may be an avenue to a promotion.

So you think you deserve a promotion? OK, so you might want to first look at your current situation closely and have all your facts and figures together. Prepare a list of your accomplishments. Are they noteworthy, undeniable, and obvious to your supervision? Is your employer in an overall growth mode and profitable? Is the timing right to make it known to your supervisor that you are not content and perhaps may even be looking for employment elsewhere?

Is it a few months before budgets are being decided or at a time when you are asked to take on responsibilities outside your current job description? If yes, then maybe it is time to present your case for promotion in a positive and upbeat way. Consider writing a memorandum to your boss and list accomplishments such as the following:

- I have never failed to meet my objectives on my assignments;
- My performance reviews have been excellent;
- I now have two patents applied for;
- I prepared and presented two technical papers at ASME conferences;
- I received special training in stress analysis;
- My salary is near the top of the grade;
- I have served as a lead engineer on most of my projects.

Many companies particularly like General Electric have a system whereby open positions are posted on company websites/bulletin boards in various departments. This

Technical Career Survival Handbook. http://dx.doi.org/10.1016/B978-0-12-809372-6.00091-8

provides a transparent opportunity for employees to make their interest in advancement known.

Caution: Be certain your current manager is aware of your application for the new position at the outset.

Song: "Give Me a Reason"
As popularized by: Tracy Chapman
1995

Demotions

There are two main circumstances for which you might experience a demotion, both of which are rare and unfortunate.

First, the overall business goes south. This may occur if the economy sours, competition takes a significant share of the market, prices of raw materials rise, or poor management decisions result in reduced profits. As an employee, you may be frustrated in this situation because you have no control over the outcome.

Second, employee performance fails to meet the goals and objectives previously established by management. This may result from failing to meet completion dates, exceeding operating expense budgets, product field failures, or disappointing clients for various reasons.

So as a technical employee caught up in these scenarios, what can you expect as a result? I have witnessed the first circumstance mentioned earlier. During the mid-1980s, there was a significant recession in the oil and gas industry, which had a major effect on SFH profits. Almost overnight, major refineries reduced their spending on spare parts for equipment they had previously purchased from us. Since spare part sales represented approximately 60% of our profits, we recognized an immediate problem.

I experienced a rather grim Monday morning staff meeting during which we were told to resubmit our department budgets with a 10% reduction in spending in preparation for a very lean year. For the initial budget rework, I indicated how I would demote personnel and reduce salaries in addition to cutting operating expenses. My plan was rejected by the GM. That is when I became familiar with the GM's philosophy regarding business declines, apparently he or she had been there before, I had not.

His philosophy was that if you demote staff, you have simply created an additional problem, a very disgruntled group of employees who in affect would be toxic. His preferred approach was to lay off staff and perhaps even provide normal pay increases for those that remained on board. Then if and when the business improved, he would consider hiring back those that were still available. Much to my surprise, I found out he was very serious when I was informed by our vice president that I was laid off effective immediately. On the plus side, I did receive a generous severance package including health care and the use of my company car for the next 6 months. Lesson learned, when the business activity turns down, determine plan "B."

Song: "Blue Suede Shoes"
As popularized by: Carl Perkins
1955

Technical Career Survival Handbook. http://dx.doi.org/10.1016/B978-0-12-809372-6.00092-X

Layoffs

In the previous section, I described the very disappointing situation when the business turns down and is forced to lay off both technical and overhead personnel to reduce operating expenses. But by reducing personnel and expenses, the business also reduces its capability to compete in the marketplace. In some instances, it may be an industry-wide situation such as the recession in the oil and gas industry that I mentioned previously. This may cause personnel to change industries and perhaps take a financial hit as a result.

So what action should you take when faced with a similar scenario? Be prepared! There are several actions I mentioned in previous sections that will help you prepare for this disastrous situation and go on offense. Think of it as a new beginning. Here are a few recommended actions you can take.

- Anticipate—If you see a turndown coming, it may be time to look for a new job.
- Employment file—Be certain it is up to date and includes examples of your latest projects and reports.
- Networking—Contact others that you know in your specialty and discipline who might provide you with job leads.
- Resume—Be certain it is current and have personnel references available when requested.
- Know your industry—Be aware of your industry and perhaps there are companies that are not experiencing a downturn and in fact, may welcome employees from their competition.
- Know relevant employment staffing firms—Be certain you have contact information for recruiters and employment agencies that specialize in your field.

Surprisingly, I used my layoff situation to start a technical recruiting (headhunter) business in a geographic area where we desired to live. I do not necessarily recommend such a drastic survival mode but you also may have had a secret desire to do something radically different. Hopefully, you will be fortunate like I was and have a severance package, which will allow you to explore new possibilities without having to worry about day-to-day living expenses for a set time.

I have seen former coworkers depart from the hazardous/radioactive (haz/rad) waste processing engineering business to deep diving equipment manufacturing, to bacteria manufacturing, and to package shipping business all because they experienced a major reboot. Another friend was laid off at DuPont, worked in the consulting field for several years then was later rehired by DuPont and regained his previous seniority.

When layoffs occur, it is expected that employers will provide a severance package to those affected to minimize the financial hardship incurred. The package provides pay as well as certain benefits when he or she leaves employment involuntarily. Depending on the financial health of the employer, the package may include the following:

- compensation based on years of service with the employer;
- employment assistance searching for a new position;

Technical Career Survival Handbook. http://dx.doi.org/10.1016/B978-0-12-809372-6.00093-1

- extended medical, dental, and life insurance;
- payment for unused vacation and sick leave;
- extended use of company vehicles;
- use of office communications equipment;
- company stock disposition.

In rare instances, these packages are offered to employees that are fired or resign. Also, when the employee agrees to accept the severance package, he/she agrees contractually not to sue the employer for improper dismissal.

Lesson learned, stay on offense. Look for signs in the company's health to anticipate layoffs or other unusual company changes. This will give you a better opportunity to survive.

Song: "Have You Ever Seen the Rain"
As popularized by: Credence Clearwater Revival
1971

Resigning

Many technical personnel think there is nothing worse than being faced with looking for a new job and are reluctant to do so even if getting fired or laid off is inevitable. I recommend staying on offense. Most importantly, do not resign until you have a very specific plan for your future. Develop your plan while you are employed. There is a simple explanation for this rule. *It is a lot easier to find a job when you have a job.* Otherwise you will need to explain why you were fired, laid off, or quit, all of which raise a question about your value to a future employer. Whereas if you are interviewing while employed, the main question becomes, what are you looking for in your next position? The answer to that question should be very clear to you.

In previous sections, I discussed being prepared, interviewing, geographics, and other topics that should help when you are considering resignation. Additionally, be certain that you will not miss a major milestone by resigning prematurely. Pensions, stock options, and employee savings plans are often tied to specific milestones such as 3, 5, or 10 years of continuous employment with your employer. Consider whether to leave your employer's qualified 401 K or 403 B with your former employer or to roll the money into an individual retirement account with your new employer based on fees, investment choices, and ability to access your funds. Your financial advisor can help you make decisions that are right for your situation especially if you have a family.

Additionally, do not presume you can remain in your position for two or more weeks after you submit your resignation. Some companies/supervisors may be generous in that regard and hope you will stay on until you finish a particular assignment and not leave them in a bind. Others may be upset and request you leave immediately. In any case, it is preferred that you do not burn any bridges. I have seen employees return to work for a former employer months after resigning from that same employer to pursue a goal that ultimately did not materialize. These so-called boomerang employees are on the rise in this current economy.

A letter of resignation should be submitted stating a concise reason for resigning but do not provide a lengthy list of gripes. Additionally, you should ask to meet with your boss face to face. Oh, yes, be prepared on how you will respond to a counter offer that I described in a previous section. It is best to be cordial and claim you are pursuing a more challenging opportunity, etc. Do not trash the company or the job thus burning bridges.

Be certain to inform your colleagues about the basis for your decision to quit and maintain contact with those whose business relationship you value. You never know when you may need to contact them as a reference or seek their help in your new position.

Technical Career Survival Handbook. http://dx.doi.org/10.1016/B978-0-12-809372-6.00094-3

Finally, consider requesting a letter of recommendation from your supervisor or perhaps a senior member of the staff who was aware of your contribution. As I mentioned in the contractor section, it is always helpful to have someone else's written assessment of your performance when searching for a new position.

Song: "We Gotta Get Outa This Place"
As popularized by: The Animals
1966

Relocation Packages

95

When engineers, scientists, or technicians are requested by their employers to relocate to an area where they will begin a new assignment, this benefit package is often provided. The range of benefits varies widely with the company, the economy and employee's salary classification and many aspects of the package may be negotiable. The relocation package may also be used in part by an employer as a recruiting package and as an incentive to transfer. According to a survey by Allied Van Lines in 2012, the relocation packages ranged from $11,000 to over $33,000.

There are typically up to five components in a standard relocation package as follows:

First, employers generally allow one or more trips to the new location for the purpose of selecting the future home. Reimbursable expenses for the trip may include transportation, meals, lodging, and possible child care.

Second, a generous relocation package may include house closing costs and during periods of high interest rates, fees to buy down the interest rates. This was common during the 1980s. Costs associated with selling an existing home may be negotiable.

Third, moving and temporary storage expenses are covered in most relocation packages. These expenses may include packing, unpacking, insurance for possessions, and shipping cars. Storage for several months to accommodate the move-in date is often underwritten by employers.

Fourth, often temporary living expenses that occur during travel and prior to moving in to a new home may be covered. Expenses such as mileage, food, and hotel or apartments may be covered until such time as the new home is occupied.

Fifth, miscellaneous expenses help reduce the strain of relocating. They may help pay for cleaning services, car license and registration, and other fees.

When negotiating a starting salary for a new position, this is the best time to request the applicable company relocation policy to understand how it will impact your decision to accept or decline the offer.

Song: "The Way"
As popularized by: Fastball
1998

Technical Career Survival Handbook. http://dx.doi.org/10.1016/B978-0-12-809372-6.00095-5

Part V

The Career

The Earning Years

96

At this stage of your career, you are hopefully well trained, experienced, and compensated. You have no doubt in experienced turbulence in the past and perhaps even some radical changes in industries or specialties. More than likely, you have been promoted to a position where you are a high-level individual contributor or in a supervisory situation. Consequently, you may need to adjust your behavior and interaction with your former peers. Here are a few things to keep in mind in that regard:

- Your work will be judged by management based on how well your subordinates perform to established goals. You are now the boss and your former peers may become the employees that complain about you and the company.
- Meet with your subordinates one on one, discuss your new role, solicit their support, discuss their goals, training needs, and obstacles they are dealing with. If you were promoted over one of your subordinates, assure him/her that you will work with them and look for opportunities for them in the future.
- You may need to disconnect somewhat from your subordinates on nonwork-related, social activities particularly happy hours and related social media.
- Do not duck opportunities to praise subordinate's performance when merited. Conversely, also be certain to point out deficiencies and the need for corrective action where appropriate.

On the positive side, at this stage of your career, you should be realizing compensation around or exceeding six figures depending on your discipline, industry, and specialty (See Fig. 30.1 of Chapter 30).

Caution: At this point, radical changes in your career path should be very carefully evaluated before taking any action.

Song: "Money"
As popularized by: Pink Floyd
1973

Technical Career Survival Handbook. http://dx.doi.org/10.1016/B978-0-12-809372-6.00096-7

Success

97

When I was contemplating an engineering career during my high school years, I had a limited knowledge of what lay ahead academically and on the job. But at that time, my father was of the opinion that engineers, scientists, and technicians realized great salary and benefits but most importantly job security. That may have been true at that time but is definitely not the case today. So if you are thinking that *longevity on the job* is a measure of success, think again. However, I have worked with engineers and technicians that were content to do the same job year after year and not complain or request a change in responsibilities. If job security is truly important to you, develop on your own. How is that done? Become indispensable (to a degree) by acquiring specialties that your company cannot do without. Evaluate the skills your peers possess and develop a strategy to acquire the skills that are essential but maybe lacking in your organization.

On the other hand, those that become bored easily like me will measure their success by "experimenting" and *transitioning to new industries and specialties* in a seamless manner and achieving praise for their performance. Additionally, to accomplish this while living in an area that offered a desirable lifestyle is the ultimate experience. I have always found that positions located in less than desirable locations came with above average compensation packages particularly those positions that were short term in duration. Companies will typically overcompensate those who make sacrifices in lifestyle and job duration.

Another measure of success for technical personnel is *creativity*. How is this measured? Like the painter, actor, writer, sculpture find satisfaction and success in their work, so too does the creative technical employee. Sure, major accolades may not always be there, never the less, creative technical personnel will realize success. This will be achieved by having the courage to be different and being dedicated to long hours and hard work. The work output of creative technical personnel may manifest itself in inventions, calculations, methods, technical publications, and manufacturing methods and procedures. The supervisor of creative personnel must be certain not to *manage* these individuals by applying the same standards applied to the more *conventional* members of his or her staff.

Success is often measured by *compensation*. If you are just now contemplating an engineering, scientific, or technician career, check current salary surveys. You will find the supply and demand economic principles apply. That is, salaries are the highest for graduates in the fields with the fewest job opportunities and vice versa. In an early section, I described the stair step salary structure. Some would measure success by their ability to move through the salary structure smoothly, year after year. If extreme compensation is your objective, you may need to think about starting your own consulting business but then be prepared for extreme turbulence. Be certain you have a solid business plan and cash reserves to cover the lean/down times.

Technical Career Survival Handbook. http://dx.doi.org/10.1016/B978-0-12-809372-6.00097-9

Think your *contribution* is minor and does not matter? During my first job as a designer at HSD, I thought my contribution was insignificant compared to the overall production of a product like the Boeing 747. Then I realized that if I had not designed the "three-wheel machine" rotating shaft assembly, the plane would not have had an Environmental Control System and therefore be unable to fly. You may be a part of a large team but their success depends greatly on your individual contribution.

Finally, *career enjoyment* may be your measure of success. For example, maybe you had an opportunity to help develop a state-of-the-art voice recognition system, helped to design a landmark football stadium or a hybrid automobile. While I was at HSD, the moon landing back pack design team was located on the floor below my aircraft component group. I remember how excited some of my friends in that group were to see the moon landing personnel on TV outfitted with those packs. If you enjoy travel, perhaps you landed a position as a technician working in an exotic vacation area. I have a friend Bill Welsh who finds it particularly satisfying to provide onsite consulting services to power plants under construction all over the world, many in exotic locations.

Song: "Take It to the Limit"
As popularized by: Eagles
1973

Retirement/Winding Down

I always thought when you did not like your work, your boss, or your company, assuming you had a reasonable income source, you retired. Now that I am of retirement age, my view on that has changed. There are some encouraging trends in this regard that I have observed are explained as follows:

Go cold turkey—Those technical personnel that I have known who were the recipients of a generous retirement "package" from their employers like General Electric, DuPont, IBM, and United Technology were the most likely to accept it and never look back. One friend, John Berube, was offered early retirement at IBM, which included credit for his years in the military, so at 51 years of age he accepted. However, not everyone makes the cold turkey transition easily. Some define themselves by their work, profession, and stature. Others put their workplace skills to work as a mentor, volunteer, or just finding new interests.

Work well past the normal retirement age—If you are healthy and not interested in pursuing other interests, hang in there. According to a recent survey by Gallup-Healthways Well-Being Index, people who work past 65 are happier than their fully retired peers unless they have no choice but to work. The job can provide both physical and social activity as a good antidote to an unhealthy sedentary and lonely lifestyle. Also, continuing to work past 65 years old usually pays off with a high salary or labor rate, particularly if you are working as a contractor. You can receive your monthly retirement payments in addition to your paycheck. Furthermore, if you continue to work, you might also start taking social security payments. This combination might produce the highest monthly income of all of your working years! A few years of "triple-dipping" could be a very big boost to your retirement funding for that point when you finally choose to exit the workforce for good.

Phased retirement—Retirees with part-time or temporary jobs have fewer health issues than those who stop working primarily when staying in the same occupation according to a 2009 study by the *Journal of Occupation Health Psychology*. Cutting back on hours but still earning an income is a growing trend to transition into full retirement. Perhaps, a sluggish economy also contributes to that trend because people prefer to avoid retiring and/or drawing down savings. Reduced hours may have a significant negative impact on benefits like insurance, pension, long-term disability, and the 401K. I recall several years ago, I was doing well in the stock market, and felt like it was time to end full-time employment. But realizing the market would likely take a hit soon after I resigned, I made a call to a consulting company that I had worked for briefly in the past and they offered me a part-time position. Shortly thereafter, I made a seamless transition into phased retirement.

Consulting—An engineer friend in the nuclear power industry saw an opportunity arise to continue working for his or her employer Dominion Virginia Power in a consulting position after retiring. The employer did insist him not to work more than 1000 h per year and paid him on an hourly basis. Shortly after, he was able to acquire additional engineering project work with another nuclear design company on a part-time basis. In this way, no career switch was required and adapting to a totally new environment was avoided. Consulting work can also be a great part-time gig for phased retirement.

Career change—During a severe economic downturn, many technical personnel may have no choice but to consider alternate career paths during their preretirement years. But going into retirement and adapting to a new work environment can be stressful according to research by Canada's Laurier University. However, one slight exception is the situation whereby the individual is transitioning to a hobby turned business. Or in the case of technical salesman friend of mine, helping his wife with her quilting hobby turned into a business.

From my perspective, most retirees I know profess to enjoy their lifestyle. Our culture puts great emphasis on careers but hopefully we will recognize when it is time to ramp down.

Song: "Should I Stay Or Should I Go Now"
As popularized by: The Clash
1982

Parting Advice

My father's favorite expression (advice), when I was a college student, was "hit the books." Well that was pretty good advice but was easier said than done. There was obviously advice imbedded in the previous sections of this book. But here is a summary of the top 10 recommendations for career survival you may have already gleaned from the previous topics:

1. **Get a technical degree**—According to Forbes magazine, individuals with engineering degrees make more money and experience less unemployment than other majors. Engineers and scientists can expect to make upward of $2.5–$3.5 million over their working careers. Technical careers are also challenging, exciting, worthy of respect, and the job outlook is positive.

2. **Continue your education**—Grow in your discipline/industry/specialty by enrolling in graduate school, training programs, and special courses that will enhance your resume. Those that hold a BS degree can expect to earn approximately 25% more than those with an AA degree. Similarly, holders of an MS degree can expect to make approximately 10% more than those with a BS degree. Read and file technical articles, start a library, cultivate mentors in your field, and you will have state-of-the-art skills desired by employers. Take an active role in a professional society and be well informed.

3. **Do not quit unless you have a new job**—Carefully prepare if you plan to resign from your job. Remember, it is easier to find a job when you have one. Employers will automatically assume you are more valuable than the candidate that is unemployed.

4. **Have a passion for work**—Engineering work can be challenging and you will have the opportunity to learn and grow on-the-job. If you love what you do, it will be obvious to your supervision and the money will likely follow. Who is more likely to get the "axe," the hard worker or the slacker?

5. **Find a job you enjoy**—Technical positions can be fun whether you are working in design, development, service, training, applications, sales, or other areas. You will develop long-term relationships when you are working side by side with other technical personnel with similar backgrounds and interests.

6. **Do not get fired**—Understand your boss's modus operandi and find out how he likes to communicate: email, phone, or face to face. Seek feedback from your boss, coworkers, subordinates, and even your customers to obtain informal input on your performance. Try to be fireproof to avoid explaining your unemployment status on that section of your next job application.

7. **Act like a consultant**—Assume you are a consultant in your field. Imagine you are in business for yourself and an expert in your specialty albeit you may be employed by a company full time. Be the go-to person in your group without being arrogant. The skills you acquire are investments in your future.

8. **Seek opportunities**—Do not be a "job hopper" but stay informed of the big picture in the technical world and seek opportunities where you fit, pursue your passion, and do what you love. Do what it takes to get the job but be professional and do not come across as desperate.

Technical Career Survival Handbook. http://dx.doi.org/10.1016/B978-0-12-809372-6.00099-2

9. **Take some risks**—Changing jobs may require some extreme relocation situations, industry changes, and unfamiliar environments. Plan your moves carefully and use a decision analysis method to weigh the factors. Do not be complacent.

10. **Do not burn bridges**—If you resign, be certain that you do not leave your projects/position/company in a bind? Review your current projects with management just before you exit and let them know they can contact you anytime with questions as they arise.

Song: "Don't Worry, Be Happy"
As popularized by: Bobby McFerrin
1988

Glossary

A.S. is an associates of science degree.

Accreditation Board for Engineering and Technology (ABET) accredits education programs at institutions in countries inside and outside of the United States. These programs span computing, engineering, and engineering technology disciplines.

Aerospace engineers design, test, and supervise the production of aircraft, spacecraft, and missiles.

Aeronautical engineers design, test, and supervise the production of aircraft.

Agricultural engineers apply engineering, technology, and science principles to agriculture and biological resources.

Alliance occurs when a mutually beneficial legal agreement is formed between two or more businesses with equity, opportunity, and risk for all parties.

Alternate fuels are known as nonconventional or advanced fuels, are any materials or substances that can be used as fuels, other than conventional fuels.

Astronautical engineers design, test, and supervise the production of spacecraft and missiles.

Acquisitions occur when business units are purchased.

ANSI American National Standards Institute.

Bill of materials (BOM) is a list of part numbers, part names, descriptions, quantities, and part numbers for a one off assembly.

Biomedical engineers develop procedures and devices that solve medical and health-related problems.

Brownfield is a site formerly used as a staging and/or storage area for chemicals.

Capital Authorization Request (CAR) is prepared when monies are to be appropriated for investments in tooling, software, or machinery.

Chemical engineers apply the principles of chemistry to solve problems involving the production or use of chemicals and other products.

Civil engineering is a discipline pertaining to the design and supervise projects, i.e., construction of roads, buildings, airports, tunnels, dams, and sewage systems.

Civil engineering technology pertains to land surveying, technical writing, mathematics, and construction and design.

Code is a system of mandatory rules for the design, fabrication, and testing pertaining to certain systems and devices. Also it is any collection of computer instructions written using some human-readable computer language.

Computer engineers research, design, develop, test, and oversee the manufacture and installation of computer hardware and related components.

Computer-aided design (CAD) utilizes computer design software to replicate product and plant design configurations.

Computer-aided manufacturing (CAM) utilizes computer software for the manufacture of objects represented by the models through computer numerically controlled (CNC) machining or other automated processes.

Technical Career Survival Handbook. http://dx.doi.org/10.1016/B978-0-12-809372-6.00100-6

Computer numerically controlled (CNC) is a machining process or other automated processes that is programmed through computer software.

Contractor is not a permanent employee but is hired to perform duties similar to a company employee.

Co-op is a student that alternates time periods between on-campus academic life and on the job in industry.

Counteroffer occurs when the employee receives an offer of employment from the employer, he/she plans to quit due to an outside offer of employment.

CPU is a central processing unit, a computer.

Design patent is a patent of a new or original designs pertaining to appearance only.

Development engineering sometimes referred to as research and development (R&D) and prepare future product or product features in prototype form.

Discipline is the academic credentials such as bachelors or AA degrees pertaining an area of study such as mechanical, electrical, or civil engineering.

Doctor of Philosophy (PhD) is a degree program to prepare a student to become a scholar and discover, integrate, and apply knowledge.

Electrical engineering is a discipline pertaining to the design, develop, test different electrical equipment.

Electronics engineers design, develop, test, and supervise the production of electronic equipment and systems.

Electrical schematic (wiring diagram) is produced by the electrical engineers defining how electrical components are interconnected and include symbols, terminals, indicators, contractors, and circuit protection devices.

Engineer-in-training (EIT) is an applicant who has passed the Fundamentals of Engineering (FE) examination and has completed any one of several combinations of education, or education and experience.

Engineering science and mechanics engineers use fundamental principles to develop engineering solutions to contemporary problems in the physical and life sciences.

Engineering manager is responsible for hiring, firing, promoting, and evaluating the performance of his/her subordinates.

Environmental engineers use principles of biology and chemistry to develop solutions to environmental problems.

Environmental control system (ECS) maintains set temperature, pressure, and humidity.

Electrical engineering technology is the design, drafting, and technical skills to assist engineers.

Feasibility study provides direction as to which design or alternative is preferred based on criteria such as cost, projected sales, complexity, safety, reliability, and ease of implementation.

Fellowship is a paid position to conduct research while enrolled in college.

Field service personnel implement corrective action, trouble shoot, and make necessary repairs/modifications at the product location.

Fire protection engineers facilitate protecting people, property, and environments from the harmful effect of fire and smoke.

Fire protection engineers facilitate protecting people, property and environments from the harmful effect of fire and smoke.

GI bill is college tuition funding for returning veterans to assist them with housing, living expenses, medical needs, education, and training.

Global positioning systems (GPS) is a space satellite navigation system that provides position and time information under various weather conditions.

Grants is a college funding that is awarded to undergraduate students based on financial need.

Green engineering design and use processes and products to minimize pollutants and risks to human health and to the environment.

Greenfield is a site whereby no hazardous chemicals are present.

Green jobs combine the concern for fuel cost and availability with protecting the environment using high efficiency machinery, low emission power generation, and alternative energy sources.

Headhunter/executive recruiter places candidates with employers and in return for a fee often on a contingency basis.

Health and safety engineers prevent personnel injury and property damage by applying their knowledge of systems and human performance principles.

Hiring authority is the individual responsible for approving a new hire.

Hybrid organization sometimes referred to as a matrix organization, is created when a line organization is combined with the product/project structure on a temporary basis.

Industrial engineers determine how to use the basic production principals, people, machines, materials, information, and energy to make a product or improve a service.

Information technology pertains to project management, systems development, networking, programming language, and the Internet.

Information technology networking pertains to networking: local area networking and wide area networking.

Internet technology refers to IT, computers, and related product manufacturing, which is generally related to principles of science.

Internships are short-term work experiences alternating with academic life.

Joint venture (JV) is a mutually beneficial business agreement in which the parties agree to develop, for a finite time, a new diverse business unit by contributing equity.

Key employee receives a unique set of company benefits and compensation.

Lead engineer is an employee who provides work direction for his/her subordinates but does not have input to evaluating their performance.

Line organization has levels of authority or seniority vertically (chain of command) and simultaneously shows peer rankings horizontally.

Magnetic resonance imaging (MRI) is a noninvasive medical test that used by physicians to diagnose and treat medical conditions.

Manufacturing centers are geographic areas where an unusually large number of manufacturers are located.

Manufacturing engineers produce mechanical and electrical components, automate assembly processes, and provide materials to the factory.

Manufacturing engineering technology pertains to automated manufacturing and materials handling using computers to design and manufacture products.

Marine engineers and naval architects design, construct, and maintain ships, boats, and associated equipment.

Materials (metallurgical) engineers develop, process, and test materials used to manufacture products.

Mechanical engineering is a discipline pertaining to research, design, develop, produce, and test tools, engines, and other mechanical devices.

Mechanical engineering technology is the application of physical principles and current technological developments to the creation of machinery and operation.

Mergers occur when business units are combined.

Mining and geological engineers find, extract, and prepare coal, metals, and minerals used by utilities and manufacturing industries.

NEC is the National Electrical Code.

NEMA is the National Electrical Manufacturers Association.

Nondisclosure agreement (NDA) also referred to as a confidentiality agreement (CA), confidential disclosure agreement (CDA), proprietary information agreement (PIA), or secrecy agreement (SA) are *legal contracts* between the employee and the company that outlines confidential material, knowledge, or data that the parties wish to share with one another restricted access to or by third parties.

Nuclear Regulatory Commission (NRC) is a governmental agency responsible for the safety of nuclear power production and other civilian uses of nuclear materials.

Nuclear engineers research and develop processes, instruments, and systems used for nuclear energy and radiation.

One line diagram is produced by the electrical engineering department to represent a three-phase power systems in block diagram format and include symbols such as circuit breakers, transformers, capacitors, bus bars, and conductor (wire) sizes.

Patent claims are carefully worded descriptions of the elements of an invention that are to be protected by the patent.

Patent search is a review of existing patents contained in the United States Patent and Trademark Office (USPTO) to determine if there is infringement on any previously patented designs.

Pell Grant is a funding method that is usually awarded to undergraduate students who have not already earned a bachelor's or a professional degree.

Pension plan provides an income source or annuity for employees in retirement after a specific term of service with the company.

Petroleum engineers design procedures for extracting oil and gas from deposits found below the earth or ocean.

Plant and instrument diagram (P&ID) is a diagram that includes mechanical, electrical, and chemical engineering information such as flow direction, piping sizes, valves open or closed, controls instrumentation, vessels, nozzles and sizes, heat tracing, and equipment and instrument numbers.

Plant engineering personnel maintain the operation of power plants, manufacturing establishments, and chemical processing plants.

Prior art is not necessarily a patent-protected design but a configuration that exists in the marketplace.

Private companies are held or owned by private investors, the company founder(s), family, or a private management company.

Process flow diagram (PFD) is a schematic that defines a system in a block diagram format showing equipment, instrumentation, piping, valves and includes operating data such as design flow rates, pressures, and temperatures.

Professional development hour (PDH) is one contact hour of instruction, presentation, or study used in the engineering community.

Product engineering group (aka commercial engineering) designs a production version of the product or a family of products that can be manufactured in quantities anticipated for the market.

Product/project organization members are focused as a team on a single product or project.

Professional engineer is licensed by a state and sufficiently knowledgeable, capable, and fit to render a design that could affect the health, safety, and welfare of the public satisfactory.

Project or product manager guides other engineers and technicians toward the completion of a project or product development program within an established schedule and budget.

Proposal is a document consisting of a scope of work, price, terms, and conditions used to render a bid on a project or product.

Public companies are owned by stockholders who purchased an initial public offering or purchased existing stock via the stock market that was previously held by stockholders.

Quality control ensures product conformance to established standards.

Reverse engineering is a process whereby a device is disassembled, measured, and studied for possible recreation.

Scholarship is college tuition funding by private organizations, churches, employers, and institutions.

Specialty is the focus within a discipline and the industry such as analysis, design, quality control, and testing.

Specification is prepared by engineers, scientists, and technicians to describe materials, components, systems, or equipment.

Spin-offs occur when business units are separated.

Standards are an established uniform engineering or technical criteria, methods, calculation procedure, test, calibration, processes, or practices.

STEM is an acronym for science, technology, engineering, and mathematics.

Superfund was legislation that was passed in 1980 to mandate cleanup of sites in the United States that were considered contaminated with hazardous and/or radioactive chemicals.

Supervisor is an employee who provides work direction for his/her subordinates and may have input to evaluating their performance.

Technical which is generally related to principles of science.

Technical degree a BS, MS, AS, PhD in science, engineering, or technology from an approved, recognized college or university related to principles of science.

Technical recruiters also called headhunters, place candidates in full time, and long-term positions in industry for a contingency fee.

Technical spectrum is a range of specialty skills that require behavior from less introverted to more extroverted behavior.

Technology refers to IT, computers, and related product manufacturing or which is generally related to principles of science.

Technology degrees are AA degrees obtained in 2 years often at community colleges or online programs related to principles of science.

Temporary employee is not a permanent employee but is hired to perform duties similar to a company employee for a specific, short-term assignment.

Trade secret is information such as formulas, patterns, device, method technique, or process used in a business that provides an economic advantage over competitors who do not know.

Utility patent is an invention that is new and nonobvious pertaining to machines, processes, or composition of matter.

Wheel organization has a manager at the center who delegates work to a small group of specialists within a company.

Further Reading

[1] 5th generation Intel Core processors. Intel; March 2015. http://www.intel.com/content/www/us/en/prrocessors/core/core-17processor.html.

[2] Occupational Outlook Handbook, architectural and engineering occupation. Bureau of Labor Statistics; January 2014. http://www.bls.gov/ooh/architecture-and-engineering/home.htm.

[3] Garnham D. 10 types of scientists. Science Council; June 2015. http://www.sciencecouncil.org/10-types-scientist.org/10-types-scientist.

[4] Top fifty jobs. Money Magazine; November 2010. p. 91–2.

[5] Department of Professional Employees, AFL-CIO. http://dpeaflcio.org/professionals/professionals-in-the-workplace/scientists-and-engineers/; December 2014.

[6] Technology Associate's Degrees. Quinstreet, Inc.; July 2010. http://www.getdegrees.com/d/associate-degrees/technology.

[7] Occupational Outlook Handbook. August 2015. http://www.bls.gov/ooh/architecture-and-engineering/mechanical-engineering-technicians.htm.

[8] Mechanical Engineering Technology. Old Dominion University; February 2014. http://ww2.eng.odu.edu/et/academics/met.shtml.

[9] Busting the 5 myths of college costs. Money Magazine; September 2013. p. 96–101.

[10] US engineering school rankings 2012. April 2015. University report. http://www.universityreport.net/us-engineering-school-rankings-2012.

[11] Money's best colleges 2015–16. Money Magazine; August 2015.

[12] Kapsidelis K. Tuition increase averages 6% in VA. Richmond Times Dispatch; August 2015.

[13] Engineering AS. J. Sargeant Reynolds Community College; January 2014. http://reynolds.edu/ciriculum/EngineeringAS.aspx.

[14] Engineering vs. Engineering Technology. ABET; January 2014. http://www.abet.org/engineering-vs-engineering-technology.

[15] Engineering Professional Education. Purdue University; January 2014. http://engineering.purdue.edu/ProEd/credit/msme.

[16] Time, Forest. What is the starting salary for a mechanical engineer with a masters degree. Houst Chronical; January 2015.

[17] Highest paying jobs: 2013. Money Magazine; September 2014. http://www.money-zine.com.

[18] FPE Graduate Program. A. James Clark School of Engineering; January 2015. www.enfp.umd.edu/grad.

[19] Engineering Career Resources & Employer Relations. Students, co-ops and internships. Penn. State University; Feb 2014. http://www.engr.psu.edu/vareer/students/coopintern.aspx.

[20] Lucas T. Interns-might be more costly than you think, legal briefs. Richmond Times Dispatch; August 2015.

[21] About ASME scholarships & how to apply. American Society of Mechanical Engineers; December 2014. www.asme.org/career-education/scholarships-and-grants/scholarships/asme-scholarships-how-to-apply.

[22] Rosen S. Schools offering bonuses to help students graduate early, on time. Richmond Times Dispatch; May 2015.

[23] Grants. December 2014. www.scholarships.com/finacial-aid/grants.

[24] Grants and scholarships, office of the U.S. Department of Education. http://www.studentaid.ed.gov/types/grants-scholarships; December 2014.

[25] Top 15 questions about the post-9/11 GI Bill. Veterans Administration; November 2013. www.benefits.va.gov.

[26] Otani A. For some employees, benefits include college. Richmond Times Dispatch; June 2015.

[27] Rosen S. Working during college can expand experience. Richmond Times Dispatch; August 2015.

[28] A better mousetrap. Site Selection; June 2014. http://www.siteselection.com/issues/2011/nov/cover.cfm.

[29] Americancitieswheremanufacturingisbooming,24/7WallSt.March2012.http://247wallst.com/special-report/2012/03/14/American-cities-where-manufacturing-is-booming.

[30] Recruiting services for employers. Virginia Tech Division of Student Affairs; July 2014. http://www.career.vt.edu/Employer/OnCampusInterviewing.html.

[31] Sidebottom J. Are you recruiting the best candidates?. Consulting-Specifying Engineer Magazine; November 2014. p. 21.

[32] Rozgus A. Wanted female eng. Consulting-Specifying Engineer Magazine; November 2013. p. 9.

[33] Did you know: demographics on technical women. National Center for Women Information Technology; July 2014. http://www.ncwit.org/blog/did-you-know-demographics-technical-women.

[34] Fidelman M. Here's the real reason there are not more women in technology. June 2012. http://www.forbes.com/sites/markfidelman/2012/06/05/heres-the-real-reason-there-arenot-more-woomen-in-technology.

[35] Landvar LC. Disparities in STEM employment. U.S. Census Bureau; September 2013.

[36] Yoder BL. Engineering by the numbers. American Society for Engineering Education; November 2014. https://www.asee.org/papers-and-publications/publications/14_11-47.pdf.

[37] Christiansen B. Women wanted for technical fields. Herald Media, Utah Valley University; June 2013.

[38] Graham A. Female engineers should get creative. Consulting-Specifying Engineer magazine; January 2014. p. 64.

[39] Careers. Money Magazine; December 2013. p. 72.

[40] Poehler G. To be a great boss, there are six traits you have to possess. Richmond Times Dispatch; November 2014.

[41] Bortz D. Shifting from buddy to boss. Money Magazine; January 2015. p. 35.

[42] Bortz D. Good ways to deal with bad bosses. Money Magazine; August 2014. p. 30.

[43] Wortman LA. Effective management for engineers and scientists. John Wiley & Sons; 1981. p. 162–76.

[44] Shanley A. Striking out on their own: independent contractors in the CPI. Chemical Engineering Magazine; June 1999. p. 92–7.

[45] Balje OE. Turbomachines, a guide to design, selection and theory. John Wiley & Sons; 1982.

[46] Management, scientific and technical consulting services. Bureau of Labor Statistics; August 2010. http://www.bls.gov/oco/cg/cgs037.htm.

[47] John Suzukida PE. Want to be your own boss?. Consulting-Specifying Engineer Magazine; October 2014. p. 19.

[48] Rivera P. Be your own boss. Richmond Times Dispatch; August 2014.

[49] Earnings, Table 1. Earnings distribution by engineering specialty. Bureau of Labor Statistics; August 2010. http://www.bls.gov/oco/ocos027.htm.

[50] Bennett J. Where the recovery isn't happening. Wall Str J October 2014.

[51] Your boss wants to keep you. Money Magazine; December 2014. p. 72.

[52] Radack DV. Understanding of confidentiality agreement, the minerals. Metals & Materials Society; 1994.

[53] The differences between small, medium and large firms. Consulting-Specifying Engineer magazine; August 2014. p. 19.

[54] Small business. September 2014. Wikipedia. http://en.wikipedia.org/wiki/Small_business.

[55] Beyer L. The rise and fall of employer-sponsored pension plans, workforce. January 2012. http://www.workforce.com/articles/the-rise-and-fall-of-employee-sponsored-pension-plan.

[56] Brandon E. Top companies continue to drop pensions. U.S. News; Oct 2012. http://money.usnews.com/money/blogs/planning-to-retire/2012/10/26/top-companies-continue-to-drop-pensions.

[57] Fields M. The 5 biggest hurdles in manager training. Orion Commercial Services, LLC; January 2014. http://www.orioncre.com/news/5-biggest-hurdles-in-manager-training.

[58] Rozgus A. MEP giants continue to diversify. Consulting-Specifying Magazine; August 2014. p. 36–8.

[59] Top employee benefits. Money-Zine; September 2014. www.money-zine.com/career-development/finding-a-job/top-employee-benefits/.

[60] Associated Press. Employers dangling unusual benefits to hook skilled workers. Richmond Times Dispatch; August 2015.

[61] Characteristics of a corporation, CliffsNotes, Houghton, Mifflin. Harcourt; September 2014. http://cliffsnotes.com/more-subjects/accounting/accounting-principles-ii/corporations/characteristics-of-a-c.

[62] What's the difference between publicly- and private-held companies?. Investopedia; September 2014. www.investopedia.com/ask/amswers/162.asp.

[63] An outline of the history of the transistor. Public Broadcasting Service; September 2014. http://www.pbs.org/transistor/album1/index/html.

[64] Measuring green jobs. Bureau of Labor Statistics; August 2010. http://www.bls.gov/green/home.htm.

[65] Hsu T. It's a green dreams gold rush. Fredericksburg Free Lance-Star; October 2010. p. D8.

[66] History of the clean air act, environmental protection agency. August 2013. www.epa.gov/air/caa/amendments.html.

[67] Wial H. Interactive: locating American manufacturing. Brookings Institution; May 2012. http://www.brookings.edu/research/interactives/manufacturing-interactive.

[68] The 2014 industry week 50 best U.S. manufacturers, industry week. September 2014. http://www.industry. .week.com/resources/iw50best/2014.

[69] Rozgus A. Four trends to watch in 2015. Consulting-Specifying Engineer magazine; December 2014. p. 7.

[70] Plant engineer: salary, duties, requirements and outlook. January 2015. http://education-portal.com/articles/Plant_Engineer_Salary_Duties_Requirements_and Outlook.

[71] Certified plant engineer, association for facilities engineering. January 2015. www.afe. org/certifications/cpe.cfm.

[72] Rozgus A. MEP Giants revenue, mergers, diversification increase. Consulting-Specifiying Engineer; August 2015. p. 25–7.

[73] Engineers. Bureau of Labor Statistics; August 2010. http://www.bls.gov/oco/ocos027. htm.

[74] Sherk J. Government jobs: nice if you can get 'em,'. USA Today; July 2010. p. 9A.

[75] Construction surveyors. October 2014. http://federalgovernmentjobs.us/job-group/ engineering-and-architecture.html.

[76] Technical field engineer federal government jobs salaries, SimplyHired. October 2014. http://www.simplyhired.com/salaries-k-technical-field-engineer-federal-government-jobs.

[77] All professional engineering positions, 0800. U.S. Office of Personnel Management; October 2014. http://www.opm.gov/policy-data-oversight/classification-qualifications/ general-schedule.

[78] Mares. Jan, 25 differences between private sector and government managers. Power Magazine; May 2013. www.powermag.com.

[79] Keller L. The pros and cons of union jobs. October 2014. http://www.bankrate.com/ finance/personal-finance/pros-cons-union-jobs-1.aspx.

[80] Sundyne fact sheet. Sundyne; Nov 2015. www.sundynecom/About-Sundyne/Fact-Sheet.

[81] Worthington-Simpson history. Nov 2015. https://en.wikipedia.org/wiki/worthington-simpson.

[82] Bhopal India disaster. Greenpeace USA; 2014. www.greenpeace.com.

[83] Kimmons R. What is a joint venture model?. Houston Chronical; 2014.

[84] Department of Labor. Worker safety series. Oct 2014http://www.osha.gov/Publications/ OSHA3252/3252.html.

[85] Singletary M. The pros and cons of telecommuting. The Richmond Times Dispatch; August 2014. p. D4.

[86] Waterman J, Robert H. In search of excellence. Harper Collins; 1982.

[87] Resume myths that will keep you unemployed. Fox Business; September 2014. http:// www.foxbusiness.com/personel-finance/2014/09/05/7-resume-myths-that-will-keep-unemployed/?intcmp=.

[88] Pofeldt E. Does your resume make you look old?. Money Magazine; March, 2011. p. 38–9.

[89] John Suzukida PE. Top 10 things to include on your resume. Consulting-Specifying Engineer Magazine; June 2014. p. 27.

[90] Goldberg B. Get a leg up on your job search. Money Magazine; March 2010. p. 36.

[91] Top 10 online job search tips. Careerbuilder.com; March 2011. http://www.careerbuilder. com/Article/CB-703-Job-Search-Top-10-Search-Tips/.

[92] Weber L, Feintzeig R. Hiring takes longer. Wall Str J September 2014.

[93] Job vacancy durations climb to new peak, DiceHiring indicators. June 2014.

[94] Durnwirth S. Survey finds disconnect regarding hiring process. Richmond Times Dispatch; Aug 17, 2015. p. E8.

[95] The history of ASME's boiler and pressure vessel code. American Society of Mechanical Engineers; March 2011. www.asme.org/engineering-topics/articles/boilers/the-history-of-asmes-boilers-and-pressure-vessel-code.

[96] NFPA 70: National Electrical Code. National Fire Protection Association; November 2014. www.nfpa.org/codes-and-standards/documents-information-pages.

[97] OSHA law & regulations, Occupational Safety & Health Administration. November 2014. www.osha.gov/law-regs.html.

[98]   Francis PH. Principles of R & D management. AMACOM; 1977.

[99]   Thomas CM. Delayed-wiper inventor wins suit against Ford. Automotive News; October 2, 2008.

[100] Blackwell, Reid J. Inventions and innovations. Richmond Times Dispatch; February 2, 2014. p. D1.

[101] Stim R. Patent, Copyright & Trademark. 13th ed. Nolo Law for All; March 2013.

[102] Franklin RA. Inventor's marketing handbook. AAJA Publishing Co; 1989.

[103] Office of policy and external affairs: patent trade secrets. U.S. Patent and Trademark Office; February 2013. http://www.uspto.gov/ip/global/patents/ir_pat_tradesecret.jsp.

[104] Regulations Governing Architects. Professional engineers, land surveyors, certified interior designers and landscape architects. Department of Professional and Occupational Regulation; July 2010.

[105] Primavera P6-enterprise project portfolio management. Oracle; Dec 2014. www.oracle.com.

[106] Kepner CH, Tregoe BB. The rationale manager. Kepner-Tregoe, Inc.; 1976. p. 173–206.

[107] 2014 Tradeshows and events. April, 2015. www.eventsinamerica.com.

[108] Hallman R. Don't whine, and do your homework before making a request for more money. Richmond Times Dispatch; December 12, 2014. p. D1.

[109] Severance package. Dec 2014. http://en.wikipedia.org/wiki/severance_package.

[110] Guerrerro A. 8 ways to graciously quit your job. U.S. News & World Report August 2013 http://money.usnews.com/money/careers/articles/2013/08/12/9-ways-to-graciously-quit-your-job.

[111] Koogler CPA, Patricia F. Top five financial considerations when changing employers. The Village News Massanutten VA; September 2014. www.villagenewsonline.com.

[112] Bloomberg News. More employees going back to their old jobs. Richmond Times Dispatch; September 2015.

[113] Brunot T. What is the standard relocation package?. Houston Chronicle; Jan 2015. http://work.chron.com/standard-relocation-package-25166.html.

[114] Brandon E. Planning to retire—top companies continue to drop pensions. U.S. News; October 2012.

[115] Cambell N. Management side of engineering. Plant Engineering Magazine; April 2000. p. 44.

[116] Lauricella T. For some, retirement brings grief. Richmond Times Dispatch; November 21, 2014.

[117] Rosat D. Retire live happily ever after. Money Magazine; January 2015. p. 47.

[118] Yip P. Is phased retirement for you?. The Dallas Morning News; January 2010.

[119] 2006 Salary survey, plant engineering magazine. January 2017. p. 25–9.

[120] Faus M. Grads await an unknown future. Village News Massanutten VA; June 2014. p. 7. www.villagenewsonline.com.

[121] Klingensmith D. Not so good advice. Richmond Times Dispatch; September 2014. p. D1.

[122] Siu PE, Sonny K. Five reasons high school students should choose engineering. Consulting-Specifying Engineer Magazine; October 2013. p. 64.

[123] Nemko M. How to avoid getting fired. Kiplinger's; July 2006. p. 87.

Index